ACTS
of GOD

ACTS
of GOD

*The Unnatural
History of
Natural Disaster
in America*

TED STEINBERG

OXFORD
UNIVERSITY PRESS
2000

OXFORD
UNIVERSITY PRESS

Oxford   New York

Athens   Auckland   Bangkok   Bogotá   Buenos Aires   Calcutta   Cape Town   Chennai
Dar es Salaam   Delhi   Florence   Hong Kong   Istanbul   Karachi   Kuala Lumpur   Madrid
Melbourne   Mexico City   Mumbai   Nairobi   Paris   São Paulo   Singapore   Taipei   Tokyo
Warsaw   Toronto

and associated companies in

Berlin   Ibadan

Published by Oxford University Press, Inc.

198 Madison Avenue, New York, New York 10016

Oxford is a registered trademark of Oxford University Press.

Portions of chapter six were previously published in *Literature and Medicine* 15, no. 1 (Spring 1996);
a version of chapter three was previously published in *Environmental History* 2, no. 4 (Oct. 1997).

Cataloging-in-Publication Data
Steinberg, Theodore, 1961–
    Acts of God : the unnatural history of natural disaster in America / Ted Steinberg.
       p. cm.
    Includes bibliographical references.
    1. Natural disasters—United States—History. 2. Emergency management—United
    States—History. 3. Disaster relief—United States—History. I. Title.

GB5007 .S74 2000
363.34'0937—dc21

                                                                        00-057991

9 8 7 6 5 4 3 2 1
Printed in the United States of America
on acid-free paper

For the little big men in my life

Nathan and Harry

There are many scapegoats for our blunders, but the most popular one is Providence.

Mark Twain

We do not see our hand in what happens, so we call certain events melancholy accidents when they are the inevitabilities of our projects.

Stanley Cavell

CONTENTS

III
·······················

CONTAINING CALAMITY

PREFACE

....

My first real job—where I had to don a tie every morning—was with a New York brokerage firm. As a summer intern in the research department, I worked the manufactured-housing beat, keeping tabs on various companies and sizing up their stock prospects. Back in the old days, before the industry sought to freshen its image, they used to call these metal boxes mobile homes. And *mobile* always did strike me as an appropriate adjective. Skyline, Fleetwood, Liberty—the names of the companies suggested a picture of luxury and freedom. But that picture was betrayed by a harsh reality: They were an inexpensive form of housing for poor and elderly people who always seemed to wind up dumped on when high winds came along. Yet in the brokerage world, another tornado was just more fodder for jokes around the water cooler, cold-blooded humor by which we attributed disaster to some magnetism inherent in these structures. It was 1981, and it was then, I now realize—when budding stock analyst met black office-humor—that this book was born.

This study examines the causal essence of calamity, peeling back the layers of obfuscation to find out not so much *what* but *who* is most often responsible for the destruction caused by tornadoes and other "natural" disasters. In undertaking such a venture, I run the risk of perpetuating a sharp division between nature (the what) and culture (the who). Of course, no such simple dichotomy exists in this Gordian-knot world of ours. And in any case, trying to disentangle the natural from the cultural is not my main concern here. Instead, I'm more interested in how drawing a distinction between *us* and *it* and blaming the latter—nature—for calamity has become a tool used to advance various political interests in society. In other words, the consequences of viewing these disastrous events as largely the work of natural or divine forces will be my main concern. My argument is not simply that natural disasters bear a strong human component, but that those in power (politicians; federal, state, and city policymakers; and corporate leaders) have tended to view these events as purely natural in an effort to justify a set of responses that has proved both environmentally unsound, and socially, if not morally, bankrupt.

My perspective is a materialist one and will no doubt strike some readers as a bit old-fashioned in light of the postmodern approaches to history so popular today. Clearly postmodernism has had a healthy effect on the historical profession, forcing people to see the multiple ways in which various cultures impart meaning to the world. Still, there is much to be learned, I believe, from studying the history of natural disasters from the vantage point of political economy, especially given how little serious attention has been devoted to the topic.

Finally, a word on style. This book is openly argumentative. It eschews the calm, even tone—or, too often, the dull and impersonal prose—that has come to typify much historical writing during this century. If at times I lapse into what some might call polemics, I do so out of a heartfelt conviction that our culture needs a passionate, critical engagement with the past to counter the denial and apathy that inform so many aspects of the response to natural disasters in contemporary America.

I want to thank my friends David Morris and Jim O'Brien for the terrific advice they have provided over the years. It's not too much to say that without them, I'm nothing. Also backing me up on this score were Michael Black, Hayes Gladstone, and Robert Hannigan, all keen critics of my writing projects. Jonathan Sadowsky provided the kind of support, intelligence, and humor that define a new gold standard for friendship. I continue to thank my lucky stars that I crossed paths with Donald Worster, one of the few historians around willing to risk it all for the sake of moral reason.

Special thanks must also go to: Miriam Anderson, Philip Banks, George Barnum, Lauren Benton, Mike Clark, Bob Cohen, Mike Davis, Seth Fein, Elsie Finley, Dave Hammack, Richard Hirn, Andrew Hurley, John Hutchinson, Judy Kaul, Dan Kerr, Linda Kremkau, Nancy Kryz, Ken Ledford, Jan Lewis, Peter McCall, Bruce Mizrach, Carroll Pursell, Martin Reuss, Alan Rocke, Marissa Ross, Kevin Rozario, Joel Tarr, Karen Thornton, David Van Tassel, and Mike Yeager. Though we have never met, I owe a special debt to the geographer Kenneth Hewitt; anyone familiar with his writing will surely detect his influence in what follows.

My agent, Michele Rubin, never gave up on this project, even when I myself had. She definitely knows her stuffed derma. Meanwhile, Peter Ginna and John Sullivan at Oxford University Press had just the right touch, and I'm eternally grateful for their efforts to improve the clarity of my argument. Susan Day was a pleasure right down to the last en dash.

The financial support provided by the John Simon Guggenheim Foundation helped in more ways than I can possibly express here. I especially want to thank William Cronon, Morton Horwitz, James Boyd White, and Donald Worster for supporting my fellowship application.

Helen Steinberg read over the entire manuscript and tried valiantly to save me from myself. Madeline Steinberg was (and is) a constant source of emotional support. Maria Del Monaco, with her unparalleled sense of justice, is, for me, the final arbiter of all that matters in the world.

INTRODUCTION

....

Hometown Blues

In Hannibal, Missouri—Hometown, U.S.A.—Mark Twain means big business. Nineteen ninety-six brought over 600,000 camera-toting tourists to the village that defines the sentimentalized picture of nineteenth-century life and culture envisioned by so many Americans. They come to visit the famed author's boyhood home, lunch at the Mark Twain Dinette, wander Main Street, pose before the most famous picket fence in history, and, above all, indulge themselves in a fantasy of whitewashed Americana. Hannibal can seem like a huge Norman Rockwell painting come to life.

But venture beyond the kitsch of this commodified hometown and you will find a darker, more problematic picture. The problems begin with the Mississippi River, the 2,000-mile-long father of waters that Twain navigated as a riverboat captain and immortalized in prose. Floods have long been a part of life in Hannibal. The first major flood on record occurred back in 1851, when Twain was still a town resident. In the recent past, however, the problem of flooding has worsened considerably. Of the ten most severe inundations in the city's history—as measured by the depth of the Mississippi—nine have occurred since 1960.[1] Needless to say, the thought of 5,000 tons of water inside a historic residence such as Mark Twain's house—the city's central tourist attraction—could make any curator nervous.

In 1985, the U.S. Army Corps of Engineers figured out a way to safe-guard the Twain home and the rest of Hannibal's historic area. It would build a 3,000-foot floodwall that would stretch from the Mark Twain Memorial Bridge all the way to Bear Creek, slightly south of downtown. When it was finished, the wall stood 12 feet high and surrounded the city like a fort. It cost $8 million to construct, and of that amount the corps paid the vast majority; the city contributed the remainder, mostly by donating land for the project. In 1990, ground was broken. Three years later, in the spring of 1993, Hannibal's wall stood completed—and not a moment too soon.[2]

The structure was designed to withstand a 500-year flood, an event with just a fraction of a chance of happening in any given year. No one could have imagined that such an incredibly rare deluge would wind up baptizing the floodwall only a month after its completion. In this respect, the wall gave new meaning to the word *timely*. The Mississippi crested at nearly 32 feet, almost double flood-stage at Hannibal, making the 1993 flood far and away the worst in the city's history. "You can see how an event like this would cause the penning of the Bible," said city engineer Bob Williamson.[3] Yet despite all the water, the floodwall worked, and Hannibal's downtown remained dry.

Environmentalists have long argued that the building of levees and walls has actually contributed to the destructiveness of floods. Although these structures offer short-term protection, when erected on both sides of a stream they force the water level to rise during heavy rain (instead of simply spreading out over the floodplain), causing it to surge over the top of the levee and punishing the "protected" area with all the more force. But in this instance, at least, the corps stood vindicated, and the agency pointed to its success as evidence of the effectiveness of the structural response to floods.[4] Build a wall high and wide enough and it may well succeed in engineering a city out of harm's way.

For Hannibal's poor, however, the 1993 flood demonstrated a rather different lesson: that the traditional response to natural disaster does not benefit everyone equally. Back in 1962, when the corps was first asked by the city to intervene in its flood woes, the agency had proposed a much longer, more equitable wall. That project would have protected the downtown *and* the city's poorer sections—the South Side and Bear Creek Bottom.[5] But the plan was shelved in 1965 after Hannibal officials failed to muster sufficient support for it; residents objected both to its cost to the city and to its aesthetic effects. Then, in 1973, a flood so devastating that it turned Main Street into a marina inspired city leaders to ask the corps to dust off the old floodwall plan. In rekindling their support for federal intervention, however, they made clear that they were no longer interested in the corps' initial proposal.[6] They wanted instead a more budget-oriented, not to mention class-conscious structure—one that would save Mark Twain's old home and the rest of the downtown, but without the added expense of protecting the city's poor.

Predictably, Hannibal's poorest residents were drenched in almost eight feet of water in 1993, leading some victims to question the city's approach to flood control. One South Side resident, Virginia Foiles, expressed frustration at the protection offered Twain's home, while her own was left to rot underwater. "They put in a floodwall to save Mark Twain's house and all the stuff about that dead man, so I don't know why they don't help the living," she said.[7] "Hannibal is for Mark Twain and Mark Twain only," lamented Donna Pagett, also of the South Side. "They could care less about their people."[8] In the Midwest, in contrast to the East and West coasts, it is often the poor who wind up living in the floodplains, a point made by another Hannibalian, Shellia Todd. "We can't afford to live anywhere else but down here where it floods," she remarked. "It's always the poor people that get screwed."[9]

Twain, with his proclivity for social protest, would have been appalled.[10] Not only had Hannibal's boosters transformed him into a commodity (and a rather sanitized one at that); they had shored up the value of their investment with a structure emphatic in its articulation of class power. "We realize that the floodwall does not protect everyone, and that is indeed regrettable," explained Twain's hometown paper, the *Courier-Post.* "But wherever a line is drawn, someone always will be just on the wrong side of it."[11] There are indeed always winners and losers in the struggle to combat natural disaster, this much is true. But there is nothing natural or inevitable about this fact, as the paper seemed to imply. Hannibal's trial by flood was certainly no act of God.

Hannibal floodwall with Mississippi River in background (*Hannibal Courier-Post*)

And neither are the other disasters discussed in this book. While it is true that nature plays a role in causing calamities like the midwestern floods, *Acts of God* is concerned instead with the human dimension of these events. Three issues will serve as the book's main focus. First, I explore the historically contingent nature of these phenomena and the question of human complicity. Why have those in power, for example, at times denied the risks of living in seismically active areas? What role has the federal government played in subsidizing life in disaster-prone locales? How has the state failed to protect its citizens against the threat of natural calamity? Second, I examine how the attempt to restore order after a disaster often winds up justifying and thereby preserving a particular set of social relations. Whose definition of normality rules during the recovery process? Whose vision of society is at stake when nature and culture collide? Lastly, I survey attempts to rationalize these events as something beyond human control. How and why have some Americans come to view natural disasters as amoral, chance occurrences? Why have they chosen to see the calamitous effects of weather or geophysical extremes as being chiefly nature's fault, obscuring the forces that compel people to live in risky environments? These three themes correspond to the three main disciplines or subfields I explore: environmental history (concerned with the interaction between human beings and nature); social history (concerned with questions of power); and cultural history (concerned with issues of meaning and interpretation).[12]

Ultimately, this book critiques the approach to natural calamity that has dominated U.S. politics over the last century. This approach has tended to overemphasize the natural factors at play while diminishing the human, social, and economic forces central to these phenomena. Natural disasters, explains geographer Kenneth Hewitt, are often seen—even by those charged with studying them—as resulting chiefly from chance, "natural extremes, modified in detail but fortuitously by human circumstances." According to the dominant view of natural calamity, as outlined by Hewitt, these events are understood by scientists, the media, and technocrats as primarily accidents—unexpected, unpredictable happenings that are the price of doing business on this planet.[13] Seen as freak events cut off from people's everyday interactions with the environment, they are positioned outside the moral compass of our culture. As a result, no one can be held accountable for them.

This constrained vision of responsibility, this belief that such disasters stem solely from random natural forces, is tantamount to saying that they lie entirely outside human history, beyond our influence, beyond moral reason, beyond control. In truth, however, natural calamities frequently do not just happen; they are produced through a chain of human choices and natural occurrences, and in this sense they form a legitimate topic for social and historical study. Thus we should begin by considering how the inert view of the "natural" disaster came to be, how it eclipsed the idea that God had brought forth calamity in response to the evil deeds of the people themselves. Whose interests has this aseptic, natural understanding of calamity served? I argue that the tendency to see nature as the real culprit emerged in the late nineteenth century but not, as one might expect, because of better scientific knowledge (though this trend clearly played an important role later). Rather, the concept of "natural" disaster developed when those in power in disaster-stricken cities sought to normalize calamity in their quest to restore order, that is, to restore property values and the economy to their upward trajectory.

The emphasis on chaotic nature as the culprit—to the exclusion of human economic forces—has in this country influenced not just the local response to disaster, but the entire federal strategy for dealing with the problem. Centered largely on prediction and control, the U.S. government's approach has been overwhelmingly scientific and technological in orientation. The careful monitoring of the nation's atmosphere, rivers, and geological formations; the building of levees and floodwalls; the introduction of cloud seeding—these are just a few examples of how attempts to predict and control nature have remained at the heart of the nation's policy response. Few would dispute that there can be positive gains from such measures. But as J.M. Albala-Bertrand has noted, natural disasters are not simply technical matters in need of more and better engineering; they are at their core sociopolitical issues.[14] The present constellation of responses has not benefited everyone equally, as the Hannibal case demonstrates. Worse still, by recruiting an angry God or chaotic nature to their cause, those in power have been able to rationalize the economic choices that help to explain why the poor and people of color—who have largely borne the brunt of these disasters—tend to wind up in harm's way. The official response to natural disaster is profoundly dysfunctional in the sense that it has both contributed to a continuing cycle of death and destruction and also normalized the injustices of class and race.

Acts of God is not meant to provide a comprehensive treatment of America's disaster history. Instead, its subject begins in the late nineteenth century—when extremes of nature first started to carry disastrous consequences in this country—and then proceeds selectively, in a loose chronological fashion, to the present. Early chapters are centered on particularly disaster-prone parts of the country (South Carolina, California, Florida, and Missouri), while later chapters are more broad-based, geographically speaking. In effect, the book spreads out from a concern with particular locales to examine the federal government's attempt to control, predict, and provide relief in response to disasters across the nation. This book is meant to unfold like a camera lens that widens its focus in relation to the increasing reach of the state into the fields of risk-production and management.

* * *

When one reflects on the array of human forces that conspired to cause disaster along the Mississippi in 1993—the corps' construction of flood-walls, the government-sponsored levee building and farming that has led to the disappearance, for example, of over 80 percent of Missouri's wetlands—it may surprise some to learn that almost one in five Americans saw the deluge as an act of God. According to a Gallup poll, 18 percent of those surveyed agreed with the following statement: "The recent floods in the Midwest are an indication of God's judgment on the people of the United States for their sinful ways."[15]

Seeing floods, earthquakes, and storms as signs of God's displeasure is arguably one of the oldest ways of interpreting these events. Consider, for example, the words of Minister Thomas Foxcroft, writing in the aftermath of the 1727 earthquake that struck New England, a fairly intense shock felt over 75,000 square miles. Foxcroft saw the event as evidence of God's "divine power." Yet he also understood the earthquake as "a Token of *Wrath* kindled against a Place for the Wickedness of them that dwell therein."[16] For the colonists, what we now call natural disasters were events heavily laden with moral meaning. They were morality tales that the God-fearing told to one another.

Foxcroft and his flock lived in a world where nothing happened at random. It was a world ruled, explains historian Donald Hall, by "radical contingency."[17] Events such as earthquakes and floods always carried a larger, deeper meaning as manifestations of God's will. Another historian, Maxine Van de Wetering, has examined sermons written after the 1727 and 1755 New England earthquakes, the latter centered east of Cape Ann, Massachusetts, and felt over some 300,000 square miles. All the texts agreed that a "moral imbalance in human behavior" had caused the ground to shake. For these ministers, de Wetering concludes, "earthquakes, especially tragic ones, were not merely luckless occasions for the chance sufferer; they were deeply meaningful punishments and conspicuous warnings."[18]

Indeed, the tendency to derive moral lessons from geophysical extremes remained a strong current in American religious thought into the nineteenth century. When three massive earthquakes occurred along Missouri's New Madrid fault in 1811 and early 1812—unrivaled in the continental United States in severity and scope—they were interpreted by many as a sign of God's power. They also seem to have inspired those who had somehow lost their faith in God to return to the fold. James Lal

Penick notes that in the midwestern and southern states, where the quakes were felt most forcefully, membership in the Methodist church increased from 30,741 in 1811, to 45,983 in 1812. That was a stunning increase, especially when one considers that Methodist membership rose by only 1 percent in the rest of the country. Many citizens of the young republic, explains Penick, sought moral lessons in natural calamities, viewing such phenomena as signs and portents.[19]

Today, if one believes the pollsters, only about one-fifth of Americans derive such moral lessons from extremes of nature. What the remaining population thinks on the matter is unclear. Many no doubt see natural disasters as simple acts of nature, a view that reflects the increasing secularization of twentieth-century American society. To most people these events probably lack any clear moral imperative or lesson. Natural calamity has become, if you will, demoralized, except of course in the sharply confined circles of the superfaithful.

This trend toward demoralization was given a boost by the state's increasing role in rationalizing disaster. Especially in the years after the Second World War, the long arm and deep pockets of the federal government assumed an ever greater share of the costs associated with natural calamities. The government provided money to repair public facilities; it funded emergency housing and offered loans; it later paid to remove debris from private property, distributed food coupons, and, beginning in 1968, provided national flood insurance. For the most part, these changes helped to underwrite increasing development in hazardous areas. But just as important, they also worked to sever risk from space. In other words, disasters were no longer simply acute local problems. Instead, the risk associated with living in, say, a flood- or earthquake-prone area was now amortized to taxpayers across the country. And when the risk of disaster was detached from the space in which it occurred, it became much harder to point the finger of blame. Ethical responsibility, not to mention ecological literacy, suffered in a world where everyone and thus no one bore the cost of residing in a hazard zone.[20]

It is also clear that the demoralization of calamity has resulted in a new set of rhetorical opportunities for those in power. Once, the idea of invoking God in response to calamity was a strategy for eliciting moral responsibility. In the twentieth century, however, calling out God's name amounted to an abdication of moral reason. With the religiously inclined less disposed than ever to take acts of God seriously, the opportunity has arisen over the last century for some public officials to employ God-fearing language as a way—thinly veiled though it may be—of denying their own culpability for calamity. In this sense, the act of God concept has become little more than a convenient evasion.

Natural disasters have come to be seen as random, morally inert phenomena—chance events that lie beyond the control of human beings. In short, the emphasis has been on making nature the villain. When Hurricane Hugo swept over the South Carolina coast in 1989, for example, *Time* proclaimed in its headline that the "Winds of Chaos" had arrived, citing a wind speed of 150 miles per hour.[21] Most news reports used a slightly lower number, at landfall, of 135 mph. Yet according to civil engineer Peter Sparks of Clemson University, the sustained winds were actually in the neighborhood of 90 to 95 mph (*gusts* were in the 100 to 130 mph range).[22] Surely there is a connection between the effort to puff up nature's fury, and thus naturalize the disaster, and the fact that even five years after the storm virtually no action had occurred on updating building codes throughout South Carolina.[23] It turns out that since the early 1960s engineers have known about the need for proper wind-loading criteria when building in hurricane-prone locales.[24] But that standard has been largely ignored out of respect for development interests, explaining why Hugo was so destructive even though it was nowhere near the mega-hurricane that the media made it out to be.

The next time the wind kicks up and the earth starts to roar, what will we tell ourselves? Will we rise up in indignation at what nature has done to us? Or will we reflect on our own role as architects of destruction? It is how we answer these questions that will determine the future of calamity.

I
RETURN *of the*
SUPPRESSED

ONE

·······················

Last Call for
Judgment Day

In 1905, Mount Pelée erupted in Brooklyn, New York—Coney Island to be exact. No place was more calamity-ridden in turn-of-the-century America. A visitor strolling down Surf Avenue could stop in at the Galveston Flood and watch as hundreds of model buildings were swept away in a monstrous deluge. The owners of the Galveston exhibit even offered patrons free shuttle service over to the Johnstown disaster, a re-creation of the deadly 1889 storm and dam-breaking on Pennsylvania's Conemaugh River. With any luck, one could still make it to the Mount Pelée eruption, where "electric appliance, water and pictorial effect" brought Martinique's 1902 volcanic tragedy to life.[1]

Perhaps no better way existed for coming to grips with the anxiety spawned by the spate of turn-of-the-century disasters than to schedule a trial run with apocalypse. If the public's "delight for horrors" seemed insatiable, as one reporter noted, it may have been because such disaster reenactments allowed people to experience their darkest fears about calamity in a controlled environment. (The proprietors of the Mount Pelée exhibit told patrons that the 1,200-seat auditorium "is provided with eleven exits, and may be emptied in two minutes."[2]) In a world increasingly defined by standardized routine, where everyday life had become enslaved to the clock (the railroads pioneered standard time zones in 1883), where engineers like Frederick Taylor were applying stopwatches to the worker's every move in order to find the most efficient way to produce things, natural disasters instead provided, as historian John Kasson has observed, an opportunity for transcendence, for "transforming commonplace routine into the extraordinary."[3] The re-created disasters, however, went one step further. They allowed people to combat the routinization of everyday life without risking their lives in the process.

But what of the relationship between the fake disasters and the real ones? As it turns out, the emergence of natural disaster as popular culture in the late nineteenth century coincided with a new code of calamity etiquette, a new script for responding to the events themselves. At precisely the same time that ordinary Americans sought transcendence in faux calamity, the business class in cities throughout America tried to normalize events such as earthquakes by draining them, as best they could, of any larger meaning.[4] The advice doled out by city leaders and local newspapers could not have been clearer: Natural disasters simply happened, and wallowing in the spectacle of life turned upside down or prostrating oneself before God only prolonged the agony. Instead, these situations demanded a calm, disciplined response aimed at putting things back in order. The Coney Island reenactments thus served as an emotional safety valve in a world where outward expressions of disaster anxiety were no longer tolerated.

Nowhere did the business class labor harder at domesticating the response to calamity than in Charleston, South Carolina, where an earthquake in 1886 drove many of the city's blacks into the streets shouting, "Judgment Day." Desperately seeking to convince these God-fearing citizens otherwise, the city's elite instead encouraged people to see the quake as a simple fact of life, as a perfectly natural event devoid of any overarching religious meaning. It is the logic behind this attempt to normalize the calamity that is our subject in the following pages. There was, of course, no one "right" response to the seismic disaster, a point made amply clear in the struggle between whites and blacks to interpret the quake in such a way as to suit their various political and religious interests. But one thing is clear: For the concept of a natural disaster to take hold, people had to internalize their fears of calamity while forsaking their own or God's role in the destruction. To understand why natural disasters today have come to be seen as chiefly beyond human control, it is necessary to explore how nature trumped God and man in the metaphysics of causation.

DISASTER CENTRAL

If Coney Island was the capital of faux disaster in the 1890s, Charleston had long been the center of real calamity.[5] It is hard to find a place more disaster-struck. Founded in 1670 on a narrow peninsula at the confluence of the Ashley and Cooper rivers, Charleston remained disaster-free during its early years. Then, in 1686, a hurricane, "wonderfully horrid and distructive," wreaked havoc on the city, which is barely 10 feet above sea level. Smallpox erupted in 1697, killing "200 or 300 persons." The following year an earthquake rocked the town, causing a fire to break out that destroyed one-third of the settlement. In 1699, yellow fever surfaced, followed by a hurricane that ripped apart wharves and flooded streets. In 1713, another hurricane swept through, causing extensive flooding and washing ships ashore. The year 1728 brought drought, yellow fever ("multitudes" died), and yet another hurricane, which sent residents scurrying to the top floors of their homes. A fire broke out in 1740, gutting some 300 houses and reducing the merchant William Pinckney and his company, the Friendly Society for the Mutual Insuring of Homes against Fire at Charles Town, to bankruptcy. One of the fiercest hurricanes in Charleston's entire history blew ashore in September 1752, killing more than 15 people and at one point inundating the city to a depth of nine feet.

Storms struck again in 1783, 1787 (drowning 23 people), 1792, 1797, and 1800. In 1804, a hurricane swept up the coast of South Carolina, putting "a great part of Charleston under water...in some places breast deep." Seven years later came a tropical cyclone combined with a tornado; the number of people killed is unclear. Two years after that a hurricane took at least 15 lives and destroyed a newly built bridge over the Ashley River. A "violent tempest" struck the city in 1822, blowing down houses and drowning eight people. In March 1838, a fire started in a fruit store near the intersection of Beresford and King streets, obliterating an estimated $3-million worth of buildings. The year 1854 brought another hurricane, followed by yellow fever, which killed 600 people. Then came the Civil War. In 1861, while Gen. Robert E. Lee inspected the city's defenses, a fire roared through town, burning more than 500 acres. Damage was estimated at between $5 million and $8 million. There were hurricanes in 1874, 1881, and 1885, the last a storm so severe that it caused the deaths of 21 people and destroyed or damaged 90 percent of Charleston's private homes. So many trees fell that it took 10,000 cart-loads to haul the mess away.

Yet since virtually all these disasters were either weather-, disease-, or war-related, the events of 1886 proved surprising to many. Several minor earthquakes—largely unnoticed—occurred in the city that summer. But on August 31, near 10 P.M. on a humid night, what began as a barely per-ceptible tremor became a monstrous earth-shaking roar that lasted for more than half a minute. The earthquake had two epicenters: one near Rantowles, roughly 13 miles west of Charleston, and one in Woodstock, 16 miles northwest of the city. The magnitude (M) of the earthquake was later estimated to be 7.0—the largest seismic event in recorded history on the eastern seaboard.* Ten severe aftershocks rattled the Charleston area for the next month. In all, a huge expanse, equivalent to two million square miles, experienced the disturbance. Tremors were felt in such far-flung places as Boston, Milwaukee, Cuba, and even Bermuda, 1,000 miles away.[6]

Because of the time of night at which the earthquake struck, no major loss of life occurred. A few died as buildings collapsed around them, but the vast majority of deaths happened outside, where people were struck by flying debris. Mrs. Jacob Middleton lost her life when the wall of the police station on Meeting Street collapsed on her. Ainsley Robson was killed by a falling piazza on Coming Street. In all, as many as 110 Charlestonians died.[7]

Property losses, however, proved even more devastating. Most of the damage resulted from the main quake in August, which destroyed more than 12,000 chimneys and caused a total of $6 million in repair work. Twenty-five percent of the total value of all buildings in the city vanished in less than a minute. Destruction was especially bad near the intersec-

---

* The Richter scale for measuring earthquake magnitude was developed in 1935. However, in the 1980s, seismologists turned instead to a better way of measuring the size of earthquakes called seismic moment. Moment magnitude (M) takes into account area, fault offset, and the rigidity of the rupturing rocks. Unless indicated otherwise, all figures for earthquakes are based on the moment magnitude scale.

Hibernian Hall (Reproduced from Clarence E. Dutton,
***The Charleston Earthquake of August 31, 1886* [1890])**

tion of Broad and Meeting streets, the very heart of Charleston. One of the city's most famous and beloved structures, St. Michael's Church, nearly toppled; its enormous portico had been ripped from the body of the church. The city hall had bad cracks in two of its walls. The main police station had been turned into a Greek ruin, the roof and entablature caving in around its huge Doric columns. Nearby, the Greek Revival

Hibernian Hall lay in a steep pile of rubble.[8] The devastation was so spectacular that visitors descended on Charleston from all over the East Coast. And they were not disappointed. Some tourists, noted the Charleston *News and Courier*, were so "forcibly impressed" with what they saw that they elected to take an early train back home.[9]

The monstrous destruction created an incredible demand for labor, driving up wages. One report issued just a few days after the calamity noted that "there was a pressing need for laborers, a fact which nobody appreciated more than the laborers themselves.... Colored men who were lounging around the streets wanted 50 cents an hour for their services at odd jobs."[10]

In truth, the earthquake could not have come at a worse time for the city. Charleston had risen to power and commercial dominance in the eighteenth and nineteenth centuries, as cotton and rice plantations expanded throughout the hinterlands. An excellent system of navigable rivers penetrating the interior allowed crops to enter Charleston, after which they were shipped out of the city's well-protected harbor. But in the years after the Civil War, the port's commercial prospects began to wane. In the 1880s, railroads crisscrossed the region, bringing crops overland to markets in the north and eclipsing Charleston altogether. When the quake struck, the city's commercial empire—once dominant over the stretch from Savannah to Wilmington, North Carolina—had seen better days.[11]

To make matters worse, the city's highest official, Mayor William Courtenay, was out of town. Interviewed at his hotel in New York, Courtenay, who had learned by this time that his wife and family were safe, was nearly speechless. "I am overwhelmed with anxiety," he said. "How to meet the exigencies of the hour, I am not now prepared to say."[12] But his own personal angst aside, Courtenay quickly collected himself and prepared to confront his ruined city. Surprisingly, one of his most formidable challenges involved not so much overseeing the rebuilding process, as calming the nerves of the city's black working class, whose unruly behavior in the wake of the catastrophe was beginning to worry his white constituents.

HIGH ANXIETY

In the eyes of the white business class, the only thing more shocking than the earthquake was the way Charleston's blacks responded to it, jamming the streets and turning the city into a kind of makeshift Coney Island.

Two days after the quake, as the sun went down on Washington Square—itself a ragtag collection of boards, canvases, old carpets, and anything else that could serve as shelter—a large group of "half-grown negroes, men and women, began to play base ball, and for a time took complete possession of the place, rushing about indiscriminately over every one and filling the air with shouts and curses."[13]

Although many of the blacks who occupied the square were, as one press report noted, "respectable people," a contingent of poorer blacks, from the "lowest classes" and "slums," were making life unbearable for the well behaved of either race. The day following the earthquake two black men "were preaching and exciting the colored women to frenzy.... They had the assurance of numbers and refused to desist, or to conduct their exercises in moderation."[14] On the night of September 3, a very severe aftershock—the strongest since the initial quake in August—rocked the city. "When the shock occurred the colored people on Washington square, as usual, became panic-stricken, and the shouting and singing was once more resumed." Eventually, such rowdiness forced white people from the site. Several days after the earthquake, the *News and Courier* reported that

> Washington square has been completely occupied by the colored people, who seem to enjoy camping out as they would a picnic.... At night the religious orgies of the colored people were so boister- ous and maddening that many of the white people were unable to stand it and were driven off, preferring to risk their lives in their ruined houses to undergoing the tortures to which they were sub- jected by these exercises.[15]

To appreciate the full meaning of the blacks' reported behavior and how threatening it was to Charleston's whites at the time requires some understanding of the ways in which race operated in this southern city. Explains historian Don Doyle, "The single overwhelming characteristic of race relations in Charleston was that a privileged white minority held onto its precarious position amid a sea of impoverished blacks."[16] This situation became even more pronounced after the Civil War, as freed-men—mostly poor agricultural workers from the Sea Islands—flowed into the city in enormous numbers. Between 1860 and 1880, Charleston's black population swelled from roughly 16,000 to more than 25,000. In 1880, blacks made up more than half of the city's population. In the postwar years, the sight of large numbers of blacks on the streets became a source of repulsion for Charleston's whites. "We very rarely go out, the streets are so niggery and Yankees so numerous," one woman wrote not long after the war ended.[17] The Fourth of July was especially galling to whites, who stayed home while blacks dominated the festivities, probably much as they did public space after the earthquake. Such behavior was not an overt act of resistance the way a race riot or strike would have been, but it might be construed, all the same, as an act of symbolic opposition. In this sense, the dramatic intervention of nature in 1886 allowed blacks to appropriate public space and to disrupt everyday social norms and behavioral codes.

In part, such transgressive behavior flowed from the blacks' interpretation of the earthquake calamity as an act of God. As Gabriel Manigault, the son of a wealthy rice planter and himself a museum curator, explained, the city's black citizens were more unnerved by the quake than were white people such as himself, who understood it as resulting from natural forces. He wrote:

> There was a marked difference in the degree of alarm which was exhibited by the whites and blacks during the first night. The former, although extremely terrified, yielded very little to any outward exhibition of emotion, and seemed to regard the event as belonging to the order of nature, while the latter were absorbed in prayer during the continuance of the minor shocks, and under the belief that this was a punishment visited upon them for their sins.[18]

A reporter covering the event for the *Baltimore Sun* also chastised blacks for viewing the disaster as a sign of God's will. It was nonsense, he opined, to understand the earthquake as divine punishment worthy of fear and frenzy:

> I find that the demoralization which is abject exists chiefly among the ignorant, of whom the large mass are colored people, who attribute the visitation to the wrath of God for the sins of the people, and not to the order of nature. The more intelligent people are busy in trying to figure out the cause of the disturbances and the chances of their subsiding.[19]

Meanwhile, the *New York Times* noted that the scientists sent to Charleston to determine the cause of the quake were "indignant because

Charleston street scene (Western Reserve Historical Society)

the ministers of the Gospel in the stricken city did not take the scientific view of the terrible demonstration of nature in their sermons last Sunday." The paper went on to say that for clergymen to see the disaster as a form of "Divine punishment" for sin "was one of those errors of judgment which do the cause of religion much harm among thinking men." It noted that "correspondents in the stricken city have pictured vividly the strange sights among the encampments of the negros, where all devote themselves to praying and shouting the weird hymns of their race." But the paper concluded that once "the terror has subsided and the city has become calm the people will begin to take that purely impersonal view of the matter."[20] Such an "impersonal view" blamed nature for the destruction, marking the shift away from divine judgment and toward the emotionally inert natural disaster.

Was the earthquake truly an anxiety-provoking act of God for black Charlestonians? It is difficult to say, but there were certainly plenty of reports that made this claim.[21] Even blacks outside Charleston, if white newspaper accounts can be believed, responded apprehensively. In St. Andrew's Parish, blacks "thought that the Judgment Day had come, and commenced crying and praying for mercy."[22] In two of Spartanburg's black churches, Baptist and Methodist, "the wildest scenes took place." Frightened churchgoers shouted and shrieked, "Judgment day" and "Prepare for judgment." In Hampton, "the colored people believed that Judgment day had come."[23] Reports also continued to highlight the differing responses of the two races. In Marion, the blacks were "wild with excitement," while the whites were "awe-stricken."[24]

Such reportage may say more about what whites thought than about how blacks behaved, but it's unlikely that the news stories were complete fabrications. Although black accounts of the disaster remain scarce, we do know that a group of prominent black ministers remained so impressed by the way "God in His wisdom has caused our city to be shaken to its foundations" that they petitioned the city council to set aside a day for fasting to commemorate the "visitation." Fast days were a relatively common response to disaster until well into the eighteenth century. But in 1886, the city council refused the request.[25]

At least one bona fide black account of the disaster does exist. Norman Bascom Sterrett, a minister in nearby Summerville, South Carolina, close to one of the earthquake's epicenters, admitted being quite terrified. "I have no apology to make for my fright," he wrote, "for every one present was frightened, and I believe I am frightened yet." Sterrett also noted, however, that in Summerville at least, both blacks *and* whites had assembled in the wake of the disaster, "mingling their voices together in supplicating the Great Ruler of all events to spare us from such a horrible death, which, without his interposition, seemed inevitable."[26]

Contrast Sterrett's view of events with that of a white bookkeeper, J.B. Gadsden. Gadsden wrote that the religious meetings of "the ignorant coloured people, as they huddled together in their terror, were among the most annoying circumstances we had to contend with." But according to Gadsden, "the agonized cries of these poor terror stricken people...did not extend to the more intelligent class; for to my own knowledge all of our servants were ready, and did their work cheerfully and well."[27] Clearly not all blacks had the same panic-stricken response. The point, however, is that in Gadsden's view, seeing the disaster as an act of God and becoming excited only interfered with the reestablishment of a normal work routine, a natural preoccupation of the white business class.

The obsessive focus on black emotionalism is perhaps the defining rhetorical feature of the white postdisaster reaction. And yet, there was nothing terribly unusual about such a demonstrative approach on the part of the region's blacks. Indeed, it would have been odd if they had *not* behaved in such a fashion. It was well-known, certainly among the white populace of Charleston and in the Low Country generally, that working-class blacks who belonged to Baptist and Methodist churches practiced a very emotionally rich form of worship. This expressive outpouring—which included singing, dancing, and shouting—was a creative adaptation of West African religion to the new theological environment of the Carolinas. In other words, such behavior constituted a form of cultural adaptation that allowed blacks to carve out a new religious identity for themselves in the South.[28] Unquestionably, whites were aware of this tendency and thus could hardly have been surprised when blacks responded to the quake as they did. Better to ask: Why were the whites so troubled by the black reaction? How exactly *should* they have behaved during the calamity?

The white Episcopal minister Anthony Toomer Porter, of the Church of the Holy Communion, offered an answer. In a letter to his "Christian Brethren among the Colored Clergy," Porter pleaded with them to have their people pack up their tents, leave the streets, and return to their homes. The sight of "able-bodied" blacks living out-of-doors when they were capable of moving inside, he reasoned, might cause those donating aid to the city to rethink their contribution. You must also stop these wild religious scenes with their loud praying and singing, he advised. "God is not deaf, and I don't suppose all the congregations are, and need not be 'hollered' at so." Continued behavior of this sort would only further divide the races, as whites came to look on such people as nothing but "savages." In place of this frenzied religious activity, Porter instead prescribed bed rest, so that one would be fresh for work the next day. As he put it: "what your people want, as do our people, is absolute rest after the excitement incident to the unusual scenes they have witnessed—I mean rest at night—that they may work steadily in the day."[29] Undoubtedly, the earthquake had empowered blacks, however briefly, to escape from the normal routines of wage labor, something that struck at the heart of white economic interests in the Jim Crow South.[30]

Although Porter agreed that God was behind all things that happened on earth, he thought it wrong to see the calamity as the punishment for human wickedness. The disaster was instead simply the result of Charleston's particular location—the product, in short, of amoral nature.[31] Porter's colleague, the Rev. Robert Wilson of the St. Luke's Episcopal Church, also criticized those who understood the disaster as an act of God. As he put it, "the man who calls this a 'visitation of God's wrath for sin' is a fanatic who ought to be silenced."[32] Porter and Wilson thus became spokesmen for the official city position, an unsurprising development given that the Episcopal church was the principal religious affiliation of Charleston's commercial elite.

Nor is it surprising that Mayor Courtenay, who returned to the city in early September, shared a similar view of the need to quell all religious frenzy. A former cotton factor and conservative Democrat who was elected in 1879, Courtenay sought to modernize the city by putting its finances in order, improving public services, and, more to the point here, helping to impose the work-discipline needed to create the proper business climate. Immediately after the earthquake, he observed, "our colored population, whose very natures are emotional," resorted to song and prayer.[33] But as he explained elsewhere, the best route to recovery in the face of the disaster was for people to get back to work. The future of the city, he wrote, "is based on *work*, not idleness, and I call upon every one to seek *work* in any and every way possible."[34] Courtney also advised the city's citizens to escape the streets immediately and return to their homes: "What our people want is relief, immediate, permanent relief from the terrible nervous strain to which they have been suddenly subjected, and which will certainly continue in the tent life which many are leading in the streets and public squares."[35] It was but one step from the streets to perhaps leaving town altogether, a point that must have severely troubled

Courtenay and the rest of the white business class, who were certainly cognizant of the fact that mobility was a central form of black working-class protest.[36]

The connection between the need for people to go back to work and the tendency to see the disaster as a "natural" event devoid of God's hand was nicely captured in the work of Francis Dawson, the editor of the highly influential *News and Courier*. Dawson, a major figure in state politics, played a leading role in helping Charleston recover from the disaster—serving on key committees and soliciting outside financial aid. Initially, Dawson was at home on the third floor of his stately house, which, he observed, had "seemed literally to turn on its axis" during the quake.[37] The force of the earthquake sent dust and debris flying and ruptured a water tank, causing a flood. But none of this seems to have deterred Dawson in the least. Within an hour, he had made his way to the office to oversee publication of a shortened edition of the paper. On the following day, however, no paper appeared, as Dawson came into class conflict with his compositors, too few of whom reported for work. This prompted him to criticize the men for their cowardice. Why they failed to report is not entirely clear. Dawson (who admitted later that he may have been too quick to condemn the men) begrudged them for thinking "more of running to their families than sticking to their posts."[38] If they and other members of the white working class were off attending the open-air religious meetings, no evidence survives to confirm it. In any case, such an obsessive concern with work, despite all odds, was in fact quite in character for Dawson (and other men of his station), who had an abiding commitment to encouraging economic progress. Under his leadership, the *News and Courier* had long sought to spur Charlestonians to shed their plantation heritage and preindustrial ways and move forward to a new economic order—the New South—with a new understanding of work-discipline and a new vision of calamity as well.[39]

With Dawson at the helm, the paper, in an editorial titled "A 'Visitation,'" spelled out the problems involved in seeing the calamity as an act of God. According to the paper, "a prominent divine in a distant city, speaking of our recent disaster, mentioned the pleasure it gave him to hear that most of our awe-stricken people recognized their trouble as inflicted directly by the hand of God. He was glad to know that in such an hour scientific theories had no value." It was the "humblest class" of people who tended to hold such a view of the calamity. Yet had the "reverend gentlemen" had the opportunity to live through the disaster and experience such "hysterical, inconsequential piety," he might feel differently about the merit of the act of God view: "Not only would he find it unmeaning in itself, but absolutely dangerous to the welfare of others." Moreover, seeing the calamity as a form of divine punishment was "entirely opposed to anything like helpful labor." As the paper continued:

> People who spend most of their nights shouting and exhorting at "experience" meetings have little strength and no inclination for work next day. To assist the community in trouble by taking up a daily routine of honest industry seems useless to them, while they are constantly looking out for the devil in his own proper person horns, hoofs and all.

It was better instead to interpret the calamity as simply an impersonal natural event, devoid of any overriding moral meaning. That way one could stay calm, get plenty of rest, and be ready for work the next day. "To work, we are told, is to pray, and this together with courage, duty and discipline, form better watch-words than 'Down in the dust;' 'It is the wrath of God!' 'A visitation.'"[40]

A LITTLE SUGAR PLEASE

Back in the 1820s, city authorities in Charleston had a bright idea. They installed a treadmill in the city's workhouse where slaves were sent for "a little sugar," meaning a whipping. In the "Sugar House," black slaves had their arms strung up above them and were forced to keep up with the treadmill while being flogged. The idea was to make slaves suffer pains above those related to their labor, so that work and docility would seem preferable to life on the mill.[41] This practice of using suffering as a tool to rein in blacks did not disappear with the end of the slave system and the rise of the free market in labor. It reared its head again in the relief campaign that the Charleston business class organized to get the working poor back on the job after the earthquake. Under capitalism, explains historian Wilbert Jenkins, Charleston's "blacks were free to work, if they could find work, and free to starve if they could not. The whip of slavery was being replaced with the lash of hunger in an economy that could provide few subsistence-level jobs."[42]

Since the entire idea of relief aid threatened the maintenance of work-discipline, many city leaders would have preferred not to have had to raise and dispense such funds. Soliciting outside aid was especially problematic because it could be construed as a sign of weakness and thereby serve to undermine a city's self-reliant image. Indeed, in 1885, when a powerful hurricane leveled Charleston and caused perhaps as much as $2 million in losses, Mayor Courtenay declined offers of outside assistance proffered by mayors from Columbia, South Carolina, and elsewhere. Cities commonly offered one another financial support in the years before the federal government became a major provider of relief. Newspapers, too, would often establish relief funds and solicit contributions from readers. But since, again, accepting such money could compromise a city's rugged, self-reliant image, some urban leaders were willing to risk the possibility of additional suffering. In his decision to decline aid after the 1885 hurricane, Courtenay had support from Dawson and his *News and Courier*, which tried desperately to inform the world that Charleston, despite the disaster, was on its feet and ready for business. Words of "discouragement and dismay," wrote the paper, would only have convinced "the commercial public...that Charleston was so injured as to be incapable of handling properly the business which was wont to be confided to her." As the paper concluded, "there must always be some suffering in a city of about 60,000 inhabitants."[43]

The economic impact of the 1886 earthquake, on the other hand, was far too devastating for Charleston's business class to consider turning away outside financial support. But very close attention was paid to the effect on work-discipline of distributing rations and other forms of aid. Charity was not "intended for those who can work," warned one relief official.[44] "The [relief] committee wish it distinctly understood that none but the actually needy and those incapable of self-support need apply for aid, as there is abundance of work, at extra pay, to be gotten in Charleston."[45]

In a calculating move, Courtenay and Dawson, who dominated the relief effort, actually cut off subsistence relief just a month after the disaster. As Courtenay explained, food relief was ended because of the "great abuses" that developed as black farm workers came from far and wide, "to the great neglect of their crops, in order to get 'free rations.'"[46] The relief committee appointed by the mayor justified the decision to halt the aid on the grounds that "normal life" had been restored, adding that should the committee be mistaken in its assessment of the situation, it "preferred to err, by limiting the estimate rather than placing it too high."[47] Nothing better demonstrates the way in which the food supply was used as a tool in the service of restoring normal, everyday patterns of work-discipline.

A month after subsistence relief ended, the black pastor W. H. Heard reported still plenty of misery in the city, at least among his own congregants: "The condition of the people beggars description—no fire, save a little out doors, poorly clad and living in damp districts. The death rate is nearly double."[48]

Not content to simply withhold immediate aid, those managing the relief effort also saw to it that only the most persistent would receive funds to rebuild their homes. Anyone seeking reconstruction money, for example, was required to complete a daunting three-page application. "If every blank in the form is not filled up, the application will be returned," would-be applicants were warned.[49] The complexity of the application drew criticism from around the country. In a letter to the editor appearing in the *New York Herald*, one observer noted that "in the history of the world there never has been a more prohibitory system of 'red tape' imposed upon applying beneficiaries where a public charity fund was to be disbursed."[50] It was a good point. By our own late-twentieth-century standards for paperwork, the application, though a nuisance, would probably deter few people truly in need. But the form was required of people who were, as yet, innocent of federal income tax and health insurance and the mountain of paperwork generated by the bureaucratic state. To make matters worse, more than four-fifths of black South Carolinians age 21 or older were unable to write.[51] None of this stopped the *News and Courier* from defending the relief committee's application process. The paper noted that it is "the duty of the committee to see to it that no one who is undeserving shall obtain relief.... It is with this view, and to this end, that the searching questions contained in the form of application have been framed."[52] In the end, the bulk of the burden for repairing buildings fell on property owners themselves. Only 2,200 of the close to 8,000 buildings in need of repair received funds from the city's relief committee.[53]

If Charleston's leaders were unwilling to share their resources with the city's poor, it was because such stinginess fit in with their interpretation of calamity. For the business class, the earthquake disaster constituted not an act of God, but a natural event and an obstacle to economic progress. The concept of an act of God implied that something was wrong, that people had sinned and must now pay for their errors. But the idea of natural disaster may have implicitly suggested the reverse, that something was right, that the prevailing system of social and economic relations was functioning just fine. No elaborate morality tales need be proffered in the aftermath of such an event, as had long been the case in the past. Instead, people were to remain calm and disciplined as they restored things to normal—effectively legitimating the prevailing social system in the process. In this view, natural disasters were not worthy of any deep or considered thought. They simply happened from time to time. Thinking about their larger meaning could only help to distract people from the task of restoring normality—with all its assumptions about the need to maximize the value of both human and natural resources. Ultimately, a view of the seismic shock as *only* a natural disaster amounted to little more than a thinly veiled attempt to return the poor back to the city's economic treadmill.

NORMAL DISASTER

One of the more "curious freaks" of the Charleston earthquake was that, for the moment at least, it stopped time.[54] Many of the city's timepieces malfunctioned in the disturbance, including the clock in the spire of the badly damaged St. Michael's Church, which came to a halt at 9:55 P.M. "The time stands recorded by the earthquake itself high up above the housetops of the city," wrote a correspondent for the *Baltimore Sun*.[55] Another report in the *News and Courier* observed that all the city's public clocks stood still at seven minutes to the hour, "as if to mark the end of time."[56]

The quake occurred at a point when Americans were still adjusting to the notion of arbitrary clock time. As historian Michael O'Malley explains, prior to industrialization, time had originated with God and expressed itself in nature—the passing of seasons and the rising and setting of the sun. Accurate time during this period was hard to come by; watches varied and even more finely crafted public clocks offered conflicting measures of the hour. But with the advent of factories, with their time-keeping technology, and the imposition of standard time to accommodate railroad travel in 1883, the clock rose to dominance. By the 1880s, as O'Malley writes, time was "expressed not in nature and seasonal tasks, but in terms of legal, industrial, and political obligations, conditions whose binding authority comes not from God, but from the interdependencies of a commercial economy."[57] Proponents of standard time encouraged Americans to simply look at the clock and go about their work rather than be distracted by deeper questions such as, "what is time and how is it set?" Similarly, as we have seen, thinking about the larger meaning of earthquakes distracted people from work, when it was better for capital if everyone went about his business. Ultimately, such efforts to streamline thinking about both disaster and time were an outgrowth of the normalizing tendencies of late-nineteenth-century industrial capitalism.[58]

With respect to the normalization of disaster, how representative was Charleston? Preliminary evidence suggests that the city's efforts to interpret the calamity to suit its political agenda may have been part of a broader trend dating from the 1860s. The last third of the nineteenth century was a period of fluid urban growth, with cities seeking to outcompete each other for economic dominance over a region. A disaster, however, could severely dampen business prospects, causing workers to flee (driving up wages) and discouraging investors wary of a hazard-prone locale. Treating calamity as a simple fact of life, however, suited the economic interests of urban elites by reasserting discipline over the labor force and reestablishing a place's image as a safe haven for business. For example, after an earthquake rocked San Francisco in 1868, the *Chronicle* lamented the "large number of panic-stricken individuals" who had left the city for Sacramento and Stockton, "in search of locations free from the awful visitations of the earthquake." Threatened by a mass exodus, the paper urged readers to remain calm and stay put in the face of "natural" forces. "If they would place themselves beyond the reach of the destructive agency of the great physical forces of nature, they must escape from the limits of this terrestrial universe."[59] Elsewhere, the paper wrote that "there is no good reason for panic"; the quake must simply be seen "as our installment or dividend in earthquake stock."[60] In other words, earthquakes were as normal as the ups and downs of the market, a matter-of-fact part of life that ought not to interfere with a city's commercial well-being.[61]

Whether elites in other cities linked the work-discipline issue to the "natural" view of calamity, as happened in Charleston, must await more extended research. Nevertheless, by the turn of the century, the concept of an act of God—however popular among ordinary Americans, became a much discussed issue in the American press, as a set of violent and deadly seismic disturbances rippled across the globe. When the real Mount Pelée (not the Coney Island model) erupted in 1902 it killed 29,000 people and reduced St. Pierre, Martinique, to rubble. Meanwhile, the disaster provided those in control of some U.S. newspapers with the perfect opportunity to cajole readers into abandoning the God-oriented view of calamity. Inhabited mainly by the descendants of West African slaves who once worked the island's sugar plantations, St. Pierre had been Martinique's chief port and its largest town. Now the *New-York Daily Tribune* called it the "New Pompeii." Explained the paper, only "primitive" people, such as the island's blacks, would foolishly engage in moral soul searching, interpreting the disaster as "the punishment of human wickedness." The more educated *Tribune* staff, however, had come to believe in "the unmorality of nature." Concluded the paper, "He is indeed a bold man, thinking himself in the counsels of Infinity, who presumes to say what are the judgments of God."[62]

In 1908, an earthquake destroyed Messina, Italy, killing approximately 120,000 people. In surveying the literature that emerged in the wake of the tragedy, the *Nation* found "little serious moralizing":

> We have the conventional expressions of pity and horror, we have generous responses to appeals for help; but we view the event with a scientific detachment which would have been impossible a century ago. Few persons speak of the hand of God; many, of the inexorable laws of nature. The difference indicates, we think, a far-reaching change in our attitude toward the universe.[63]

Undoubtedly, the shift in thought was the product of the growing importance of scientific reason, that is, the trend toward the secularization of culture, a process the sociologist Max Weber dubbed "the disenchantment of the world." Yet whatever the source of this more impersonal view of calamity, it was a notion that readily served the instrumental purposes of the nation's urban elites when "nature" threatened closer to home.

By the early twentieth century, as the *Nation* pointed out, people had increasingly come to see earthquakes as "*merely* natural phenomena."[64] This detached and morally neutral view defines, more than anything else, the meaning of natural disasters in modern America. More important still, the trend toward seeing these events as "merely" freak natural acts has helped to paper over the human role in them, at times depriving Americans of the chance to shape their own culture's destiny. Of course, sometimes disasters do "just happen," no matter what steps people have taken. Nevertheless, the view that chance geophysical factors are the *primary* culprit has led to the apathetic political atmosphere surrounding natural hazards policy, a point borne out, as we will see, by the failure over the last century to incorporate the full meaning of the Charleston tragedy into the East's approach to seismic preparedness.

HERE?

Given the great lengths to which Charleston's leaders went to keep the earthquake disaster from interfering with the city's commercial agenda, one imagines that they would not have been too disturbed to learn that 100 years later it had been all but forgotten—consigned to obscurity, where it could do little harm to the area's prospects for continued economic growth. In the century after the quake, construction went on, and builders for the most part were utterly unconscious of the seismic hazard, functionally reproducing the same fatalism inherent in the act of God interpretation.

As late as 1983, according to civil engineers James Nau and Ajaya Gupta, the majority of the city's buildings and facilities were "still without adequate seismic resistance." In the early 1980s, metropolitan Los Angeles had some 12,000 buildings lacking in seismic adequacy; yet Charleston, despite being a much smaller city, had even more such compromised structures.[65] Another study by the U.S. Geological Survey included interviews with emergency officials, building inspectors, architects, and engineers in Charleston regarding the area's seismic problem. "With the exception of building inspectors and structural engineers," the report observed, "few of the respondents currently incorporate awareness of an earthquake hazard into their decisions." Although nearly all those questioned were familiar with the 1886 disaster, "few of the people with whom we spoke took seriously the possibility of an occurrence of a future earthquake of the magnitude of the 1886 earthquake."[66] In this sense, the earthquake was seen as a unique event, with little meaning beyond its value as a quaint episode in the city's past, a fact mentioned on horse-and-buggy tours but otherwise consigned to the dustbin of history.

Perhaps that is why it took the city until 1981 to pass an amendment to the building code requiring adequate seismic design.[67]

To the extent that the 1886 quake was seen as a problem, it was understood for nearly a century to be Charleston's problem alone. Throughout the East, very little attention has been paid, until very recently, to the threat of earthquakes. John Lyons, the director of the engineering laboratory at the National Bureau of Standards, observed as late as 1984 that despite the lack of detailed surveys "it is fair to say that seismic design is simply not practiced" in the eastern and, for that matter, even central United States.[68] To be sure, the East is much less seismically active overall than the West.[69] But although the probability of an earthquake is lower, the risk of a major seismic calamity is actually very high. This is because seismic energy tends to travel greater distances in the East.* Compare, for instance, the 1886 earthquake with the 1971 San Fernando quake. Both were of roughly the same magnitude. But the Charleston quake made itself felt over an area 10 times as large.[70] This means that heavily populated metropolitan areas like New York City could well experience the effects of a distant seismic disturbance. Indeed, the 1886 disturbance actually cracked walls in Harlem, a point not lost on Manhattan structural engineer Guy Nordenson. Skeptical of seismic awareness in the Big Apple, where city officials procrastinated until 1995 before incorporating seismic provisions into the building code, Nordenson asked: "Who still doesn't get it? Why, everybody. You've got 8 million people in New York; how many seriously think we've got an earthquake problem?"[71]

Why have earthquakes not been seen as a problem in the East? In part, the lack of awareness stems from the failure to locate the geological formation responsible for the Charleston disaster. Unlike California, where the source of seismic activity is routinely identified—correctly or not—with the famed San Andreas fault, no such structure has been discovered in the South. As late as the early 1960s, scarcely more was known about the 1886 quake from a scientific perspective than at the time it happened.[72] Even in the 1970s, the dominant scientific view of the disaster held that unique conditions existed at Charleston, localizing the earthquake problem. Hence in its decisions regarding where to locate nuclear

---

* The more rigid the underlying rock, the less resistant it is to deforming under stress. Cooler rocks tend to be more rigid than hotter ones—commonly found in the West—and thus capable of greater seismic wave propagation.

reactors, the Atomic Energy Commission ruled that while design requirements in the Charleston area were to be in keeping with the risk of another 1886 magnitude earthquake, along the rest of the eastern seaboard the seismic history of this southern city would not influence the blueprints one bit.[73] The significance of the 1886 disaster has thus been minimized—interpreted as a singular event of only localized importance.

In truth, however, the Charleston area is not geologically unique. Many studies have tried, but none has succeeded, to prove the distinctiveness of the city's underlying geologic structure.[74] In 1982, the U.S. Geological Survey—realizing that seismicity, not geological formation, was all that distinguished Charleston—issued a clarification of its views on the significance of the 1886 earthquake. Scientists, the statement noted, had been unable to connect the disturbance to a geological structure, and thus the possibility existed, remote though it might be, of a similar disturbance at another point on the eastern seaboard.[75] In short, it stated something that had been slowly becoming obvious: Large magnitude earthquakes are possible elsewhere in the East.

The Charleston calamity, starved of its meaning for future generations, is little more than a footnote in the annals of disaster, a very poor cousin, as we shall see, to the famed San Francisco earthquake 20 years later. Cordoned off as a unique—shall we say freak—event, the Charleston experience has been ignored by developers and planners as they have forged ahead in the building of such projects as Manhattan's Battery Park City, which is constructed on a quake-prone landfill. Yet as the most deadly and destructive quake ever to strike the East, the 1886 disaster should stand as a symbol of what the denial of seismic risk might lead to, a reminder that earthquakes are not a concern for Californians alone, a warning of just how important it is that we examine the geography of risk and how it is produced and, most of all, a prediction that the coming eastern earthquake calamity—whenever it occurs—will not be an act of God.

TWO

....

Disaster as
Archetype

Twenty years later and the earth spoke again. The 1906 San Francisco earthquake is arguably the event that defines calamity in the popular imagination. It is the Big One that lurks in the back of the American mind. As the geographer Kenneth Hewitt has put it, the turn-of-the-century disaster has come to be seen through a process of "historical-geographical compression" as an icon representing the entire seismic risk problem, "even as many other disasters occur and are forgotten."[1]

Mircea Eliade has talked about the ahistorical quality of popular memory, that is, "the inability of collective memory to retain historical events...except insofar as it transforms them into archetypes."[2] The 1906 San Francisco calamity stands out as one such archetype. It exists in an interpretive void of sorts. On the one hand, it is canonized as the natural disaster to end all disasters; on the other hand, its meaning rarely transcends the realm of caricature and myth. A search of some 38 million records catalogued by U.S. libraries uncovers 572 citations under the subject. Its closest competitor in terms of the sheer volume of written, photographic, and map material is the Johnstown flood of 1889, which generated only a quarter as many items (146, to be exact).[3] Meanwhile, within San Francisco itself the disaster continues to serve as a key memory marker in the city's collective unconscious.

But for all the talk of 1906 representing the very epitome of bigness, the event is, when put into context, hardly the most sizable earthquake on record. That distinction goes to a 1960 Chilean quake (M 9.5), a full 350 times more powerful in terms of seismic energy.[4] Moreover, the notorious San Francisco quake, for all the tremendous attention lavished on this one slip of earth, has hardly had the effect on development and building in the city that one would expect. In this sense, the disaster has both tremendous meaning and almost no meaning at all, at least not in its impact on reducing seismic risk throughout the Bay Area.

SMOKE AND MIRRORS

The battle to interpret the San Francisco disaster began even before the smoke had cleared. That struggle pitted those seeking to capitalize on the disaster's entertainment value against California's business class, which expressed deep reservations about the adverse role all the publicity might have on the city's commercial prospects. Less than one week after the disaster, the *New York Times* reported the preparation of 100 "distinct and separate books telling the complete story of the San Francisco earthquake and fire."[5] In fact, at least 82 popular accounts of the calamity, often lavishly illustrated, were published in 1906 alone, an extraordinary commemorative outpouring. And that number does not include the many newspaper and magazine accounts republished in book form. Nor, of course, does it include the huge number of separate photographs and postcards that circulated throughout the country.[6] "Ever since the disaster of April 18 the cooler members of the community have looked askance at the wide dissemination of photographic views of San Francisco 'after the earthquake and fire,'" observed an editorial in the *San Francisco Call*. The paper, owned by John Spreckels, one of California's most prominent capitalists, objected to all the publicity. "Are we not damaging the city by every one of these views we send away?" The editorial continued: "The whole world is familiar with our calamity, but is it necessary to harp on the subject after it is all over? Why not forget it as soon as possible.... If we want to frighten people away from us this is about as good a way as any other."[7]

Calamity was big business in turn-of-the-century America, and obviously many publishers saw the disaster as a potential source of profit. But, needless to say, for most of San Francisco's business class, the disaster was not something they chose to advertise. At the time of the earthquake, the city led the West in trade and manufacturing and was fast becoming a major financial center as well. As far as the business class was concerned, the quake could not be allowed to impede San Francisco's commercial future. William Humphrey, president of the Tidewater Oil Company, noted that the calamity seemed to inspire "wonderful fraternalism" among the city's commercial leaders. "Everyone was in the same boat, so we forgot all else and pulled as a team."[8] And at nothing did they work harder than in shaping the way the calamity would be understood. There were great stakes involved in how the nation perceived the disaster.

The 1906 earthquake registered M 7.7, roughly five times the magnitude of the 1886 quake. It occurred along the San Andreas fault (the most visible strike-slip fault in the world) and resulted in a rupture of the earth's surface that extended more than 250 miles. Although felt as far south as Los Angeles, as far north as southern Oregon, and as far east as central Nevada, the earthquake is still commonly understood as exclusively a San Francisco calamity. In fact, the quake caused extensive damage throughout northern California.[9]

The shock occurred a little after 5 A.M. and lasted about one minute. Subsequently, fires erupted in San Francisco as electrical wires were severed and gas mains exploded. The fires burned for three days over an area of almost five square miles. Unable to stem the blaze because underground water mains had been damaged by the quake, the fire department stood by as over 28,000 buildings and residences succumbed to the flames. Annihilated were the business district, vast parts of the factory and entertainment areas, the major hotels and restaurants, as well as nearly all of the city's most important buildings.[10] Exactly how many people died remains unclear, as we will see, but 3,000 is by no means far-fetched.

Nineteen hundred and six, of course, was not the first time that the San Francisco area had been rocked by an earthquake. Geologists have found signs that a shock of comparable magnitude occurred sometime in the mid-seventeenth century.[11] Much more concrete evidence is available of an 1838 quake registering M 6.8 (intensity, not moment magnitude) along the San Andreas fault; but this disturbance caused little damage because San Francisco was then just a tiny hamlet. Thirty years later, with the population of the city nearing 150,000, another M 6.8 shock took place, this time along the Hayward fault, which runs roughly 65 miles from east of San Jose to San Pablo Bay, although only about 30 died.[12] Indeed, in the 50 years prior to the 1906 event, earthquakes ranging from M 6 to M 7 happened one or more times each decade in the Bay Area.[13]

The 1868 disturbance, in particular, stands out for the damage it caused on so-called made ground. In the 1850s, in the midst of the Gold Rush, Yerba Buena Cove was filled in and incorporated into San Francisco's business district. This area of some 200 acres experienced the brunt of the destruction when its unconsolidated soil took on the properties of quicksand, a process now known as liquefaction.[14] Although the science behind liquefaction remained unknown in the 1860s, the damage to buildings on landfill was readily apparent, and newspapers took note. The attention, however, had no real impact on future development in the city until after the Second World War, when builders began to take advantage of advances in soil engineering science.[15]

The 1868 earthquake is also notable for spawning one of the more mysterious episodes in the annals of disaster. Soon after the calamity, George Gordon, a San Francisco real estate developer, urged the city's chamber of commerce to form a committee to study how to build earthquake resistant structures. But, curiously, the committee's work was never made public. George Davidson, a scientist and committee member, later claimed that Gordon himself sabotaged the group's efforts. As he wrote in 1908, "The report was carefully prepared but Mr. Gordon declared that it would ruin the commercial prospects of San Francisco to admit the large amount of damage and the cost thereof."[16] The committee, explained Davidson, had estimated $1.5 million in damage—five times the amount that Gordon and his business associates had made public in their telegraph messages to eastern capitalists. Exactly what Gordon's motives were remain unclear, though this may not have been the only such effort to divert attention from northern California's seismic problem. "The prevailing tone in that region, at present," wrote geologist Josiah Whitney in 1869, "is that of assumed indifference to the dangers of earthquake calamities."[17]

Indeed, to hear the commercial community tell it in 1906, one would scarcely even imagine that San Francisco had a seismic history. Marion Scheitlin, who covered the disaster for the *Chicago-Record*

San Francisco, 1868 (Bancroft Library)

Herald, wrote that "one hears the dual disaster referred to most as 'the fire.' Every effort is being made to induce capital to come to the city, and it is acknowledged here that capital is more chary of earthquake than of fire." Capitalists were wary of investing in seismically active locales because at the time it was believed that, in contrast to earthquakes, more could be done to avoid fires. Realizing the threat that seismic activity posed to the city's commercial future, "the men who are devoting their time to the restoration of confidence and the rehabilitation of the city, are very certain to minimize the disastrous effects of the earthquake. The expression is heard constantly: 'There would have been but little damage had not the fire started.'"[18]

No organization was more dedicated to stoking the fire-oriented view of the disaster than the Southern Pacific Company, the dominant economic force in California at this time. Railroads were notorious in the West for their promotional activities, and when the 1906 calamity struck every effort was brought to bear in one of the great disinformation campaigns of turn-of-the-century America. James Horsburgh Jr., general passenger agent for the Southern Pacific railroad, explained that the company had no intention of "advertising the earthquake" when "the real calamity in San Francisco was undoubtedly the fire."[19] In a promotional tract titled *San Francisco Imperishable*, the company observed that the earthquake seemed to appeal to "the imaginative and emotional." However, "the main mission of this message from the Southern Pacific Company is...most emphatically, that the destruction was due to fire and not to earthquake."[20] In fact, the 1906 quake was powerful enough to damage 95 percent of the chimneys in San Francisco.[21] Going by the Southern Pacific, however, one would never have believed that a quake measuring M 7.7, among the most powerful in the history of North America, had occurred.[22]

The Southern Pacific was hardly alone in seeing fire as the main cause of the calamity. It had support from the business class throughout California, from the governor on down. Eager to show that the disaster had not been allowed to get in the way of business, Governor George Pardee observed that he had declared only 41 legal holidays. This figure compared favorably, he pointed out, with the 74 such holidays declared after the Baltimore fire, "which occasioned much less damage." "The earthquake," he continued, "severe and destructive as it was, did not do, as has been so wildly heralded, much damage, in comparison with the following fire.... It was the fire, and not the earthquake, that laid half of San Francisco low."[23] Lining up behind the governor was the secretary of the California State Board of Trade, Arthur Briggs, who reported: San Francisco "has been destroyed by fire.... The earthquake damage relatively was inconsiderable."[24] And one could certainly count on the California Promotion Committee to fan the flames. Set up to lure people into migrating to the state, the committee issued this statement intended for audiences in the East: "The earthquake did some damage to poorly constructed buildings.... The disastrous effects of the fire were appalling."[25]

The city's newspapers, which were effectively an instrument of the social order, also fell into line in an attempt to downplay and normalize

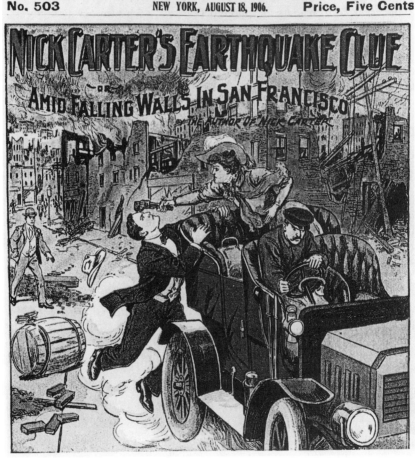

This comic book sullied the city's image by playing up the quake and its damaging effect on order. (Museum of the City of San Francisco)

the calamity. The *Chronicle* constantly reminded readers that the city must be rebuilt and that earthquakes must not get in the way. "A few have been heard to express alarm lest the fact that the fire was preceded by an earthquake might retard the growth of the city," editorialized the paper. But the quake would have no real effect whatsoever, the paper continued. "Except while the earth actually trembles, and for a short time thereafter, nobody cares anything about earthquakes.... The earthquake in this city is of no consequence, except for the distress resulting to those present in the city at the time, who will speedily forget it."[26] The *Bulletin* told its readers that more people died each summer in large eastern cities from sunstroke than lost their lives from the earthquake:

> Familiar dangers lose most of their horror. If we had a severe earthquake every year or two in San Francisco we should think as little of earthquakes as people in Kansas think of cyclones. But because this earthquake was unique, because it was a calamity of a kind which had not happened before in California, people think and talk as if it were far more terrible than flood or pestilence, fire, wind or sun."[27]

In terms of the damage caused, the 1906 disturbance was certainly unique. But this attempt to fashion the disaster as a freak occurrence was ultimately little more than an attempt to rationalize continued development by discounting the region's seismic past.

The San Francisco Real Estate Board outdid all other organizations in its attempt to pass the disaster off as no more than a fire. Just a week after the quake, the board met to discuss the calamity. "It was agreed," explained a report on their meeting, "that the calamity should be spoken of as 'the great fire,' and not as 'the great earthquake.'"[28]

Among individual members of the business class, probably the most spirited seismic denial came from Sen. Francis Newlands (D-Nevada). Newlands, whose San Francisco real estate experienced severe damage in the quake, played a critical role in securing more than $2 million in federal aid for the city. A patron of the Southern Pacific, Newlands is of course well-known for his obsession with developing the West.[29] So it is perhaps unsurprising to find him supporting a vision of the calamity consistent with continued growth in a city where he had a major financial stake. According to his account, what the city had on its hands here was a natural disaster, not an act of God. As he put it, "the forces of nature seized this great city and shook it like a rat in a trap. Nature itself contemplated no serious harm; it simply demonstrated its force by shaking the earth a little." Newlands believed the quake caused only 3 percent of the destruction, though how he arrived at that figure is unclear. Twenty percent is now the accepted number. "Earthquakes mean nothing," explained the quake-defying senator. "A little shake in the earth's crust, resulting in a crack here and there of a few inches, constitute no real source of danger. The destructive element here was the fire."[30]

On the one hand, the business class was selecting among dangers, that is, making a political decision about what San Franciscans should and should not fear—fire versus earthquake. And on the other, it was seeking to depoliticize the calamity, to drain it of meaning, to encourage people to forget the disaster in the hopes of hiding the fact that interpreting it as a fire amounted to a very political act.[31]

EARTHQUAKE COUNTRY

More than any other group, it was the scientific community that spoke out against these attempts to interpret the earthquake away. In 1909, writing in *Science*, the eminent geologist Grove Karl Gilbert observed, "It is feared that if the ground of California has a reputation for instability, the flow of immigration will be checked, capital will go elsewhere, and business activity will be impaired."[32] Some years later, John Branner of Stanford lamented the paucity of information on the 1906 calamity. He attributed this lack of knowledge to the commercial community's "deliberate suppression of news about earthquakes." As he explained:

> Shortly after the earthquake of April 1906 there was a general disposition that almost amounted to concerted action for the purpose of suppressing all mention of that catastrophe. When efforts were made by a few geologists to interest people and enterprises in the collection of information in regard to it, we were advised and even urged over and over again to gather no such information, and above all not to publish it. 'Forget it,' 'the less said, the sooner mended,' and 'there hasn't been any earthquake' were the sentiments we heard on all sides.[33]

Branner's colleague, Andrew Lawson, the man who named the San Andreas fault, also commented on the widespread efforts to soft-pedal

earthquake risk. "The commercial spirit of the people fears any discussion of earthquakes for the same reason as it taboos any mention of an
occurrence of the plague in the city of San Francisco. It believes that such
discussion will advertise California as an earthquake region and so hurt
business."[34] But at this point at least, Lawson and his colleagues were
helpless before the tectonic act of denial that rippled out across northern
California's political landscape.

Just three days after the 1906 quake, Governor Pardee appointed a
scientific commission, including Gilbert, Branner, and Lawson, to investigate the effects of the seismic shock.[35] Pardee appointed the committee,
but that was all he did; he offered the commission no money to conduct
its work. In the end, the group had to turn to a private source, the
Carnegie Institution of Washington, for the funds to publish its two-
volume report in 1908 documenting the calamitous effects of the earthquake in detail.[36]

Apart from the publishers and the scientific community, only one
other group seemed interested in taking the earthquake seriously: the
people who were going to have to foot the bill, namely the insurance
industry. Once the fires had been put out, a small army of insurance
adjusters converged on the city. They faced one main problem: to distinguish between losses caused by the earthquake exclusively—not covered
under the standard insurance contract—and those resulting from fire,
which the companies would pay for.* Obviously, most policyholders were
screaming fire because they had a vested financial stake in seeing the
destruction this way. Five weeks after the initial shock, an article in
Insurance Field reported: "Among the people with losses to adjust, the fact
that there was an earthquake has been forgotten. In fact it is now tacitly
understood that there never was any earthquake, and that the whole trouble
was the fire."[37] The product of what might be called the "insurance
parallax"—where "truth" followed economic interest all too closely—the
fire-oriented view wound up costing insurance companies across the

---

* Separate insurance coverage for earthquake risk did not exist at the time. And the so-called
  fallen building clause contained in fire insurance policies protected companies from having to
  pay damages when a building collapsed or exploded before it caught fire. However, generally
  speaking, insurance companies did cover fire losses that were the *indirect* result of the seismic
  shock. It was thus in the best interests of insurers to figure out how much damage the earthquake directly caused so as to limit their liability.

Postcard of the San Francisco fire with quake damage removed by an artist (Museum of the City of San Francisco)

world money. Frederick Hoffman, a statistician and insurance expert, wrote some years after the calamity that companies were "forced repeatedly to make payments which were not justified, as word was passed 'not to talk about an earthquake but about a terrible conflagration.'"[38]

Indeed, the pressures of land speculation in San Francisco (combined with the threat that quakes posed to such activity) proved so unrelenting that even some insurance people there felt compelled to downplay the area's seismic risk. The *Coast Review*, a San Francisco insurance periodical, responded to fears voiced in the eastern press about California's seismic past by asserting in the wake of the 1906 calamity: "California is not an earthquake country and never has been." Noting that earthquakes had occurred in Charleston and near St. Louis, the publication identified the stretch from the eastern seaboard north to the Great Lakes and as far west as Kansas as "the real earthquake country."[39] Placed in context, this claim is not quite as absurd as it might sound. With the Charleston calamity only a generation old and California's status as the site of the vast majority of U.S. earthquakes not yet common knowledge, one could at least argue that the Golden State lacked any unique seismic problem.

California, to be sure, may rightly be called earthquake country today, but this was not a predestined outcome. The Californization of seismic risk was manufactured as a product of western expansion and land development in league with twentieth-century scientific knowledge about the relatively high seismicity of the state. Nonetheless, it is striking how in San Francisco at least, attempts to shed this emerging image and deny the region's seismic hazard wound up having some very sorry consequences. After the 1906 earthquake, a local committee of the American Society of Civil Engineers urged the city to strengthen its building code.[40] In the summer following the calamity, the city followed suit and passed a new law requiring that buildings be constructed to withstand a wind force of 30 pounds per square foot. Predictably, the new law did not mention earthquakes by name. The section dealing with seismic stresses is titled "Wind Bracing," and one imagines the language here was chosen deliberately.[41] In any case, only three years later, the standard for wind bracing fell by one-third, to 20 pounds.[42] And by 1921, the standard originally recommended by the engineers had been cut in half, to just 15.[43] This downward trend led the seismologist Bailey Willis to write in 1924 that San Franciscans "have reason to anticipate the next earthquake with apprehension, for they have allowed the conditions favorable to an otherwise avoidable disaster to grow up in their midst." Willis lamented the "unsafe building conditions," which he claimed were "well known to architects, engineers, and other intelligent citizens."[44] The eminent civil engineer John Freeman, present in San Francisco on the twentieth anniversary of the 1906 calamity, also observed buildings that were being constructed "in the cheapest manner possible." City officials in San Francisco and throughout California, Freeman remarked, "appeared strangely lenient toward the speculative builder."[45]

Such lenience stemmed directly from the conspiracy of seismic silence that remained a major preoccupation of San Francisco's business community well into the 1920s. In 1925, an earthquake in Santa Barbara (M 6.3, surface wave), several hundred miles south of San Francisco, killed 12 and destroyed many brick and concrete buildings in that city's business district. Boosters flew into action, seeking to keep the disaster from sullying California's image as a safe haven for business. California, Inc., an organization based in San Francisco and set up to promote development in the state, pressured newsreel companies not to harp on the devastation. The group also met shortly after the disaster with booster organizations throughout California, including the San Francisco and Los Angeles chambers of commerce. They agreed among themselves to minimize the actual destruction and not to seek relief contributions from outside the state. Predictably, donations to Santa Barbara's relief effort failed to materialize. The California Development Association, charged with soliciting contributions from within the state, reached just half its fund-raising goal.[46]

There were, however, some signs in California of a willingness to deal openly with the state's seismic risks. Unlike the 1906 calamity, no fire followed the 1925 quake, a fact highlighting the destructive nature of the seismic shock. Several months after the disaster, the Commonwealth Club of California, a group made up of businessmen and professionals, formally recognized the threat that earthquakes posed to the northern part of the state. The group called on San Francisco, in particular, to revise its building code accordingly.[47]

As long as earthquakes remained a fictional threat in California, insurance to protect against such phenomena was a hard sell. But after the Santa Barbara calamity, which caused losses valued at roughly nine times more than the $600,000 paid out by insurance companies, earthquake insurance suddenly came into fashion.[48] The increase in earthquake coverage was actually fueled by a spate of lesser known seismic disturbances in 1925, including a severe quake (M 6.6) that struck east of Helena, Montana, just two days before the Santa Barbara calamity. Earlier in the same year, the East Coast was shaken by an earthquake (M 6.2, surface wave) centered in Canada's Charlevoix-Kamouraska region. The tremors spread out over roughly two million square miles, extending from eastern Canada as far south as Virginia and as far west as the Mississippi River. More people probably experienced this shock—it was felt in New York and Boston—than any other seismic event up until this time.[49] The country seemed in the midst of what one observer writing in the *Independent* called an "earthquake crisis," an observation suggesting that the equation of earthquakes with California alone had yet to solidify into fact.[50]

In response to nationwide seismic activity, many Americans rushed out to see their brokers. Earthquake insurance had existed since 1916, but as late as 1924 premiums totaled only a little over $200,000. The following year, stimulated by the various quakes that rocked the nation, premium income skyrocketed to over $2 million. Most of that increase in coverage, however, was on property based in California.[51] Thousands of Californians were now putting up their own hard-earned money to protect themselves against a risk that the business class had quite recently said did not exist.

What accounts for this shift toward a more open recognition of earthquakes, especially on the part of California's business class, is difficult to pinpoint. Obviously, the disasters themselves and the scientific ability to single out seismic activity alone—without the complication of fire—proved very important. It was also clear that the policy of denial did not always make good economic sense, and worse, could actually impede attempts to right society after a disaster. Since the late nineteenth century, authorities in U.S. cities had encouraged people to remain calm in the face of calamity in order to get on with the business of restoring economic life and property values to their predisaster state. But denying the possibility of calamity could thwart this goal by leaving people unprepared and creating a context in which panic could occur, a point made by one Los Angeles insurance executive in 1925.[52] In addition, by the 1920s, California was richer and more populous than ever before. Between 1900 and 1920, San Francisco's population increased from 343,000 to 507,000, while the assessed value of property nearly doubled, rising from $413 million to $820 million.[53] It was clearly self-defeating, if not foolhardy and bad business, to fail to protect such vast amounts of wealth and people against a risk that clearly existed. Finally, one wonders whether the crude attempts at denying seismic risk were wearing thin by the 1920s, as publicists moved away from such direct attempts to influence behavior toward a more indirect and nuanced psychological approach that preyed on people's unconscious yearnings for instant wealth.[54]

Then, in 1933, an earthquake (M 6.3) rocked southern California along the Newport-Inglewood fault—yet another event that helped to firm up California's status as the nation's earthquake capital. "An earthquake may be legally described as an act of God," intoned the *New Republic*, noting that 116 lives were claimed in the disturbance. But those who died were hardly the victims of "God's will," the magazine continued, calling the nation's attention to the political economy of risk in California. "It is not too much to say that many of them, if not all, were murdered—murdered by the cupidity of California business men."[55]

Most of the damage was centered in Long Beach. Again, as in Santa Barbara, no sweeping fires erupted. Again, there was no denying that an earthquake, and only an earthquake, caused the destruction. Forty million dollars in damage resulted, with school buildings faring especially poorly. With the country in the midst of the Great Depression, the state's boosters could not afford to minimize publicity about the calamity by limiting their fund-raising appeal to California alone, which had been their strategy back in 1925. Indeed, this time the state turned to the federal government for help. Five million dollars in federal relief eventually flowed into the state, but in return for this money, California's leaders were forced to concede the obvious: that a disaster of significant magnitude had, in fact, occurred.[56]

The earthquake was also notable in another respect. It led to the first major legislative initiative in California to recognize the region's seismic problem. Outrage over all the schools damaged (luckily, classes were out when the quake hit) in part moved the legislature to action. The result was the Field Act of 1933. The act imposed seismic safety standards on school buildings and was soon followed by the Riley Act, applying similar earthquake resistant criteria to all new buildings. According to historian Carl-Henry Geschwind, this new era of seismic enlightenment dawned when seismologists and engineers—motivated by a Progressive Era concern with using natural resources efficiently—marshaled enough scientific data and financial reserves (secured from businessmen, including some in San Francisco) to convince the state's leaders that economic development depended on open recognition of the earthquake risk.[57]

In 1936, on the thirtieth anniversary of the quake, the Fireman's Fund, which nearly went under in the calamity, published a report surveying three decades of seismic denial. "The lessons of 1906 were speedily forgotten," the report observed. "Public apathy, an aversion to admitting that earthquakes occur in California, and the desire of building construction speculators to build for profit, combined gradually to prevail over the counsel of engineers." San Francisco's building code was weakened. The standard for wind bracing was reduced. Floor loads were cut. "Happily, this is now becoming history, although it required another major catastrophe—southern California's earthquake of 1933—to awaken the public once more to the necessity for taking steps for its own protection."[58] If nothing else, the seismic events of 1925 and 1933 seemed to prove that California was earthquake country after all.

Now the scene was set for the 1906 earthquake to go Hollywood. It is no accident that only after California's business class ended its love affair with the ideology of denial did Metro-Goldwyn-Mayer decide to cash in on what has become one of the most famous quakes in history. MGM's *San Francisco* opened in 1936 and starred Clark Gable, Jeanette MacDonald, and Spencer Tracy. Reviewers took note of the fairly pedestrian plot (centered on a love triangle) and then gushed over the terrifying earthquake scene that capped off the film. *Esquire* wrote: "You actually feel the ground splitting under your seat."[59] And according to the *New York Times*, the earthquake scenes were so downright frightening that the actors themselves "frequently and involuntarily rushed from the sets or sought shelter outside the main stages."[60] "An earthquake in the real Metro-Goldwyn-Mayer manner," observed *Time*, "it lasts for 20 minutes on the screen and in all respects except casualties no doubt betters its original of 30 years ago."[61] Imagine a movie about the 1906 earthquake that "betters"—that is, was worse than— the real thing, a movie made by Californians themselves, no less.

The irony here was not lost on one of Hollywood's most famous film critics, Rob Wagner. So impressed by the magnitude of the faux disaster that he "looked for the nearest EXIT," Wagner wrote that his first reaction was, "What will…the California Chamber of Commerce say?" As he continued, "Here I've been telling Outlanders that our California earthquakes are a joke, that we use 'em to rock the babies to sleep, that in San Francisco it was the Fire. And now along comes MGM and not only rocks the publicity boat but capsizes it."[62]

The Californization of earthquake risk took yet another giant step forward in 1938. A map prepared by the chief geologist of California's Division of Mines showed for the first time on a state-issued document all known geologic faults. Although the Seismological Society of America had issued its own statewide fault map in 1923—a major achievement in itself—this was the first time a government mapping effort had seen fit to include information of seismic importance. The 1938 map superseded a 1916 map published by the California State Mining Bureau, which had failed to include any delineation of faults whatsoever—omitting even the San Andreas. And this, even though such geologic knowledge was available as early as 1910 in the atlas accompanying the state earthquake commission's investigation of the 1906 calamity.[63]

Seismic enlightenment was hardly one great march upward, however. Real estate interests and others seeking to boost land development in the San Francisco area have tended to revive the tacit silence on matters of earthquake vulnerability when it has suited them. Beginning in the mid-1950s and continuing on through the next decade, as the Bay Area experienced explosive growth in the demand for housing, seismic risks were again actively denied in the haste to increase real estate prospects. Writing in 1964 for the *Nation*, David Cort observed that

> the very word, earthquake, is taboo in California. This prudery does not prevail in other earthquake areas, but California's tone is set by the realtors, as well as the undertakers. Earthquake talk lowers real estate values and raises insurance rates. The old joke about the "San Francisco Fire of 1906" is familiar, but it is no joke.[64]

No joke indeed, as events in Redwood Shores in the 1960s were to prove. Located on the San Francisco peninsula, Redwood Shores was once marsh and mud before dikes were used to reclaim it. In 1963, Redwood City, in search of additional tax revenues, joined with a salt company, interested in making use of thousands of acres of unused evaporation ponds, to announce plans for a major housing development. Situated just 5 miles from the San Andreas and 14 miles from the Hayward fault, the site consisted of soil that some believed posed a major seismic hazard. The city, however, thought otherwise. It hired a well-known scientific consulting company that reported nothing terribly unique about the Redwood Shores site, nothing that made it any more seismically hazardous than parts of San Francisco itself, certainly nothing to cause anyone to think twice about building on the mud.[65] Enter G. Brent Dalrymple and Marvin Lamphere of the U.S. Geological Survey. The two voiced serious concerns before a California legislative committee about the idea of building a new real estate development on landfill, citing evidence going back to 1865 showing that it is generally more prone to seismic stress than terra firma. Although testifying unofficially as private citizens, the two geologists were chastised by their superiors in Washington and told not to answer any further questions. That happened after George McQueen, a public relations man for the real estate interests, and J. Arthur Younger of San Mateo, the district's congressman, put pressure on U.S. Geological Survey officials not to let seismic safety stand in the way of development.[66] Later, Congress, involved because the federal government was offering mortgage guarantees for the project, briefly blocked the progress of the venture, citing concerns for seismic safety, but eventually caved in and allowed the building to proceed.[67] Despite considerable evidence that landfill posed seismic hazards, including reports after the 1906 earthquake that buildings on "made ground" fared especially poorly (not to mention firm scientific evidence dating from the 1960s on liquefaction), the developers of Redwood Shores went ahead with their plan. Today, there are roughly 5,000 housing units on the landfill, waiting for the effects of another major quake to test the wisdom of the developers' decision.

THE CROWDING-OUT EFFECT

All the attention lavished on the 1906 calamity, the San Andreas fault, and California more generally, has tended to crowd out any recognition of seismic hazards elsewhere in the West. Salt Lake City, for example, is rarely mentioned as a potential seismic flash point when, in fact, the Wasatch fault runs straight through its heart.[68] Roughly 75 percent of the population of Utah lives near the fault, which has generated M 7.5 earthquakes in the past. California may well be rightly considered earthquake country, with the state accounting for the vast majority of U.S. seismic activity.[69] But one must make a distinction between the risk of earthquakes and the risk of *disaster*, a distinction that is often elided. And in this sense, California has hardly cornered the futures market in catastrophe, as developments in Utah demonstrate. In the early 1990s, state geologists there tried to upgrade the earthquake risk designation of the northern and most populous part of the state, a move that would have increased building costs, but by a trivial amount (estimated at 0.5 to 1.5 percent for new office buildings). The building industry, however, defeated the move. James Bailey, a structural engineer and building industry sympathizer, remarked, "You don't design for earthquakes that occur every 2,000 years—it's not reasonable." In fact, according to the Utah Geological Survey, an M 7.0 earthquake along the Wasatch has a recurrence interval of 350 to 500 years.[70] Again, here is another case where recourse to "freak nature" has been used to sanction more development, a move that could generate a catastrophe with loss of life and property destruction equal to anything previously seen in California.

Meanwhile, the mythic status of 1906 and the San Andreas—the most famous fault system in the world, it is often said—has also diverted attention from other immediate seismic threats in the Bay Area itself.

This is not to say that the focus on the San Andreas has not led to progress. One of the most important scientific insights concerning earthquakes—the elastic rebound theory, still valid today—resulted from the intense study of the 1906 calamity.[71] But the obsessive concern with the San Andreas has caused many to overlook the threat posed, say, by the Hayward fault, which, surprisingly, is as close to the center of San Francisco as its better-known counterpart.[72] The Hayward fault was last active in a destructive way back in 1868, an event that is now but a distant memory for most, and not by accident, as we have seen. Moreover, it hardly helped matters, as Karl Steinbrugge noted in the 1960s, that the threat of legal action on the part of property owners "reportedly kept some geologists…familiar with the location of the Hayward fault from publishing their knowledge in detail."[73] As a result, development prior to 1972 (when the Alquist-Priolo Act made seismically irresponsible building illegal) saw the siting of many structures on active traces of the Hayward zone.[74] "The way many hospitals and schools were built around here, you'd think they used the fault as a guide for where to place those buildings," said Lloyd Cluff, who in the 1970s served as an advisor to the California legislature on seismic matters.[75]

A 1987 study showed that an M 7.5 disturbance on the Hayward fault would cause 1,500 to 4,500 deaths in the Bay Area.[76] Yet the fault itself remains far from the consciousness of many who live near it, as a recent study has shown. Residents of Santa Clara County, interviewed soon after the 1989 Loma Prieta earthquake, were asked to identify the closest active fault to them. Just 21 percent of those living near the Hayward were able to correctly name it. Even more shocking, 57 percent of those interviewed misperceived the San Andreas as being closer and thus posing a more significant threat.[77] Such is the power of the world's best-known fault system that even Californians adjust their mental geography in order to accommodate it.

IMPERFECTLY SAFE

To this day, the question of just how many people died in the 1906 disaster remains unclear. Is there a connection between the way the calamity was domesticated to avoid undermining the city's commercial prospects and the failure to fully reckon the death toll? Maj. Gen. Adolphus Greeley, who commanded the Pacific Division of the army, called in to restore order after the quake, reported 498 deaths. The city's subcommittee on statistics, meanwhile (relying on the coroner's office for its information), offered a figure of 674 (confirmed dead or missing).[78] Yet Gladys Hansen, the disaster's foremost social historian, disputes both figures as far too low. In the early 1960s, while working as San Francisco's official archivist, Hansen fielded phone calls from people looking for information on relatives killed during the calamity. In answering such queries it dawned on her that the number of inquiries far surpassed the officially reported death figures. In a city with a population in excess of 400,000, struck by a monumental catastrophe, 500 to 700 dead seemed like an awfully small number. She pointed out further that Chinatown was obliterated in the calamity, while just 12 Asian names appeared on the subcommittee's official death list. Embarking on a detailed examination of the issue, Hansen compiled a database of confirmed deaths—names verified as dead from more than one source. That list now stands at 3,000, but Hansen cautions that even this figure is probably far too conservative, predicting a final death tally of over 5,000.[79] For all the attention paid to the San Francisco disaster, it is remarkable how the constant focus on property destruction has eclipsed the truly deadly story here.

Perhaps not surprisingly, it was the transient and working poor, many of whom lived in hotels in the area south of Market Street, who suffered most in the 1906 disaster. The hotels were built on an area of landfill that was once the old Mission Bay Swamp. When the earthquake occurred, a violent chain reaction ensued, as the quake caused the hotels to collapse like dominos.[80]

In 1989, when the Loma Prieta earthquake (M 7.2) struck the San Francisco Bay Area, the dangers of building on landfill were again amply demonstrated by the heavy damage sustained by the Marina district. Although less powerful than the 1906 earthquake, the 1989 quake nevertheless managed to cause buildings to collapse and water and gas mains to rupture. The Marina was once a giant lagoon before city fathers filled it in with, among other things, the burnt ruins created by the 1906 disaster (to make way for the Panama-Pacific International Exposition of 1915). Now the buried past came back to haunt residents of this tony neighborhood.[81]

The media focused obsessively on the Marina, showing residents in Docksiders hauling their belongings about in plastic trash bags. But in truth, the worst hit area in the city was precisely the same one flattened in 1906: the south of Market area that was still home to skid row. The Loma Prieta earthquake reduced 4,700 units of multifamily housing in the city to rubble, precipitating a major housing crisis among the poor. To this day, under half of the affordable rental housing destroyed in the Bay Area has been replaced.[82]

When the Loma Prieta quake hit, San Francisco had yet to pass an ordinance to deal with the city's roughly 2,000 unreinforced masonry buildings. Some 800 of these vulnerable buildings housed approximately 25,000 residents of the city's low-end rental market in Chinatown and other poor sections. Yet unlike the 1906 calamity, which seismic expert Karl Steinbrugge notes did not lead to "any long lasting improvements in earthquake resistive construction," the 1989 quake moved the city to act on the unreinforced building issue, something it had been considering as far back as the 1970s.[83] In 1992, the city's voters passed a ballot initiative creating a $350 million loan program requiring building owners to retrofit them no later than 2006. Today, many owners are still dragging their feet, waiting to see if the city is serious about enforcing the ordinance. Meanwhile, the architect Michael Johnstone wonders why the city has given landlords such an extended period of time to comply. Building owners have long expressed concern that the high cost of retrofitting would ultimately wind up forcing out tenants who can't afford higher rents. But Johnstone is skeptical. "The idea of not inconveniencing someone—but allowing them to be killed, well, things seem upside down."[84]

In fact, there was nothing illogical here, at least not by the reigning standards of our economic order. Such thinking simply reflected the monetary calculus that informed the city's retrofitting program. Calculating that $835 million worth of structural redesign would save 415 lives (given an M 7.0 earthquake on the Hayward fault), and $335 million only 235 lives, the city chose the cheaper route. "You can't make a perfectly safe world. You always have to balance costs into whatever level of safety you make," said Kathleen Harrington, president of a property owners group that supported the city's retrofitting initiative.[85] All this may well be true. But it overlooks the ultimate question: Whose lives are we talking about here? Clearly, the lives of some are worth more than others. Given the demographic make-up of the areas containing the suspect buildings, it seems almost certain that it will be the poor and people of color who will suffer the most in a future earthquake. These are the kind of people, then, who are truly at risk, unsurprisingly the same people who were buried in 1906, people who never made it onto the official death lists, people who have largely disappeared from the collective social memory.

Unable to fully embrace the meaning of the 1906 calamity, except as an archetype, unable to see the underlying political economy of risk that explains who lives and who dies when the earth shakes, northern California's real estate industry is content to mortgage the future of the region's dispossessed. Until this pattern is put to a stop, the San Francisco fire will continue to smolder.

THREE

························

Do-It-Yourself
Deathscape

"**F**lorida has its hurricanes; California, its earthquakes."[1] So reads the caption beneath a photograph of the San Francisco calamity in a 1963 geography of the Golden State. Although that may sound like an obvious insight, in fact, there is nothing natural or predestined about this state of affairs. At the time of the West Coast disaster no such statement was possible, because Florida had yet to become the magnet for hurricane calamity that it is today. The problem with such a view is that it effectively naturalizes a particular set of geographies of risk, instead of asking how these historically specific hazard profiles came to be. Florida, like California, was not born risky. It was built that way.

Why is South Florida a disaster waiting to happen? One reason is that private developers have sought to maximize the region's tourist and agricultural potential by building in areas susceptible to hurricanes and flooding. Meanwhile, the state of Florida, playing the role of accomplice, has not been able to give away land and natural resources fast enough to private interests, providing a hefty subsidy to developers who gobbled up the region's scenic but hazardous barrier islands and other places prone to recurrent inundation. Private-property-driven economic development helped to sow the seeds of future destruction, while Florida's business community sought to deny the very real risks involved and, where possible, to blame nature or God when disaster did occur. Natural disaster has a very shadowy history in Florida, rooted in years of denial for the sake of more hotels and suburban sprawl. Below we probe the sources of that denial, exploring the unnatural history of natural disaster in the state.

MAKING OF A MANSCAPE

Originally, Miami Beach was just a thin spit of sand, 200 feet wide, with a low ridge of dunes running down it. West of the dunes, toward the bay, weeds gave way to a tangled swamp of red mangroves, a species with dense roots and stems. A swamp crisscrossed by mangrove and buzzing with mosquitoes and sand flies was hardly the kind of tropical paradise attractive to sun-starved northerners. So developers launched a makeover of the landscape, transforming it into something it was not.

The man responsible for initiating this transformation was Carl Fisher, an automobile magnate best known for building the Indianapolis Speedway and the first cross-country highway. Fisher was an incredibly daring man, not at all shy about taking risks and, evidently, not a big fan of mangrove swamps. An avid racer of bicycles, boats, and cars, Fisher once staged a promotional stunt in which he donned a padded suit and rode a bicycle across a tightrope stretched 12 stories above the ground. On another occasion Fisher sailed over Indianapolis in a car that dangled from a balloon. Will Rogers, in a generous moment, said Fisher had done more "novel things" than anyone he had ever met.[2] A man seemingly desperate for adventure and risk, Fisher first came to Florida on vacation in 1910 and several years later purchased 200 acres of land on what would eventually become Miami Beach.[3]

In the summer of 1913, Fisher and several other investors hired a dredging company to clear away 1,000 acres of mangrove swamp and replace it with six million cubic yards of bay-bottom land.[4] Black laborers were hired for the brutal, insect-ridden task of slashing through the tangle of mangroves. Then a bulkhead was erected and bay bottom was pumped up to neatly cover the shore. Soil from the Everglades was later shipped in, and the land was divided into lots and planted with shrubs.

"In this manner," writes Polly Redford in *Billion-Dollar Sandbar: A Biography of Miami Beach*, "the original landscape was erased as if it had never been and a more salable one built in its place."[5] More salable, the landscape surely was; but it was also more hazard-prone now that the mangrove, which offer natural protection to coastal areas from hurricane-induced storm surges, were destroyed.

Fisher's genius, of course, was to anticipate the extraordinary premium that investors would soon place on waterfront property. In a promotional tract published in 1926, *Florida in the Making*, the authors observe what they call a "mania for water frontage." "There is nothing mysterious or deep about this buying. It simply means that in people's minds the ocean, the gulf and the lakes are Florida; and aside from the farmers, the buyers do not care so much for property that has no water frontage."[6] By draining the swamp at Miami Beach and paving it over, Fisher would soon realize huge profits in land sales. In 1925 alone, he and his business associates sold property on Miami Beach totaling over $23 million.[7]

Turning water into new land was indeed something of a compulsion for Fisher.[8] Not content simply to dredge and fill along the barrier beach itself, in 1917 he focused his attention on creating artificial islands in the middle of Biscayne Bay. In his first such venture, Fisher purchased bay bottom from the state, erected a bulkhead, and then pumped in sand to create Star Island. Later he helped to subsidize a bond that led to the building of the County (now MacArthur) Causeway, linking the island—soon to become a yacht club—with the mainland. Eventually he sold the property for $200 per waterfront foot, a fact that did not escape the calculating eyes of fellow developers. In the early 1920s, land specu-lators built bulkheads and filled the Venetian islands, linking them to the mainland via another causeway. By 1931 alone, some 6,000 acres of land had risen Atlantis-like from the depths of Biscayne Bay.[9]

Florida has a long history of handing over the state's natural resources to private enterprise. Of the approximately 35 million acres of land and water in the state, over 23 million were controlled by a state board responsible for the welfare of such lands. And all but a mere 3 mil-lion acres were deposited in private hands by the turn of the century. By the 1950s, almost all the land under the board's control had disappeared. Various pieces of legislation dating from the mid-nineteenth century helped to fuel this monstrous shift of public resources into the pockets of private entrepreneurs. The legislature became even more accommodating to real estate developers in the early twentieth century, as the value of waterfront property in South Florida became increasingly clear. A 1913 law, for example, encouraged the development of property in the middle of bays, vesting title in many submerged lands in the state board and allowing the board to allot this public resource to private interests.[10] Indeed, Carl Fisher and other speculators in the Miami area availed themselves of precisely this key piece of legislation to help turn Biscayne Bay into a miniature Venice.

If there is something to be said for creating increasing amounts of waterfront real estate—either from a developer's or tourist's perspective—it is also true that in Florida, doing so placed more people and property in harm's way. In 1915, property in Miami Beach was assessed at approximately $250,000. A decade later that number had increased to a whopping $44 million.[11] By 1926, Miami Beach, once an undeveloped barrier spit, had become a real estate theme park complete with more than 50 hotels, nearly 200 apartment houses, and over 800 private homes. For those who chose to notice them, however, signs abounded that all this property had been built on an extremely unstable piece of land. As early as 1918, before the land boom in the 1920s, enough evidence of serious erosion existed to necessitate the installation of two 60-foot jetties in front of the Roman Pools, a bathing pavilion and casino, built (like most of the hotels along the ocean) far too close to the water.[12]

Despite evidence of these and other more life-threatening hazards, developers just could not get close enough to the sea. In the spring of 1926, only months before one of South Florida's most devastating hurricanes, a group of land promoters came under fire for planning to turn a set of coral reefs south of Cape Florida into artificial islands. It was a suicidal idea that would certainly have led to death and damage had the scheme gone forward. But major opposition on the part of landowners around Biscayne Bay intervened to scuttle the plan. Typically, the majority of deaths from hurricanes result from storm surges, when low pressure causes a large dome of water to swell over the land. If a landmass is convex, the impact of the surge tends to fan out and dissipate. Biscayne Bay, however, is concave and enclosed by offshore islands. As those opposed to the real estate venture argued, let the plan to build more islands on the coral reefs proceed, preventing water from returning quickly back to the ocean, and the effects of the storm surge would surely mount, pounding the shore with all the more ferocity.[13]

That some land-hungry souls seriously entertained building on the vulnerable reefs is testament to the reckless abandon of real estate development in this period. The Florida land boom, John Kenneth Galbraith has observed, centered on a craving for instant gratification, that is, "an inordinate desire to get rich quickly with a minimum of physical effort."[14] But evidently for some, the yen for quick wealth brought with it another form of fantasy, almost a death wish.

SMILE

The Miami real estate boom went on under fair skies. By 1926, it had been two decades since the last intense hurricane had made landfall in the area. Even the October 1906 storm (category 2), which caused considerable damage and loss of life, did the vast bulk of its destruction south of the city, in the lower keys.* Seeming immunity from violent weather, coupled with the haste to cash in on the red-hot real estate market, resulted in a profusion of hastily and poorly built structures. Many homes constructed during the 1920s boom had little or no bracing and were designed without regard to wind pressure and hurricanes.[15] "Real estate buyers who inquired about the menace of tropical storms," read one newspaper report, "were laughed at."[16]

One could perhaps still hear the laughter even as late as July 1926, when a category 2 hurricane bore down on the city. After the storm ended, traffic ground to halt on the County Causeway as curiosity seekers tried to catch a glimpse of what the strong winds had done, though far more significant damage occurred further north in Palm Beach.[17] No lives were lost in Florida (although some 150 died in the Bahamas), leading the *Miami Herald* to conclude that as fearful as hurricanes may be, "there is more risk to life in venturing across a busy street."[18] Before

---

* Hurricanes are categorized according to the Saffir-Simpson Damage Potential Scale, developed in 1972. Before that time, hurricanes were divided into Minor, Major, Extreme, or Great Hurricanes, the last having winds in excess of 125 mph. The five-level Saffir-Simpson scale uses wind speed, central pressure, and storm surge heights to determine hurricane category and damage potential. Thus a category 2 storm (moderate damage) has a central pressure of between 28.50 and 28.91 inches, a maximum sustained wind speed of 96 to 110 mph, and a storm surge of 6 to 8 feet. A category 5 storm (catastrophic damage), by comparison, has a central pressure of less than 27.17 inches, a wind speed of greater than 155 mph, and a storm surge in excess of 18 feet.

Miami, 1926 (Florida State Archives)

scarcely a week had passed, meteorologist R.W. Gray of the U.S. Weather Bureau told members of the local Kiwanis club that Miamians need not fear serious damage from a hurricane.[19]

Events were to prove otherwise. On September 18, 1926, a category 4 hurricane—the twelfth most intense to make landfall in U.S. history—battered South Florida. In the midst of the raging storm, Philip London, a Miami businessman with a cameraman in tow, set off to film the disaster for the benefit of moviegoers in New York. Threading their way through debris-clogged streets, the two made it to Miami Beach and then forged their way up the coast as far as Fort Lauderdale, capturing the grisly horror of the storm. "Life was following its happy, careless course in Miami on Friday last," explained London upon arrival in New York City. Morning newspapers, he continued, had carried reports of a hurricane; afternoon papers denied the possibility: "We dismissed the thought." London slipped into bed only to be awakened at 2:30 A.M. by a jolt and a loud roar. Then the lights went out. Later in the morning, London peered outside and saw a tangle of telephone poles and wires. The Halcyon Hotel had lost its roof. A big electric sign on top of the Olympic theater had been broken in two. A little before noon, the winds let up and the rain stopped, allowing people to venture outside. "Everybody tried to smile and appear happy," London reported. The owner of a downtown store hung out a sign that read: Start the day with a smile![20]

Certainly, Florida's boosters did everything in their power to encourage such optimism. Indeed, one wonders how they felt about London's movie, which presumably sought to capitalize on the calamity's terrifying effects at the same time that Florida's business community tried to assure the world that things were not so bad after all. Shortly after the 1926 disaster, the *Literary Digest* wrote that "It has become a habit in resort regions to minimize or even suppress news of disasters that might frighten away tourists and investors." As in San Francisco, earthquakes, storms, and other calamities were "kept out of the daily papers and the dispatches of the great news-gathering agencies."[21]

Keeping adverse publicity to a minimum presented quite a challenge to Florida's business community in the decade following the 1926 storm. In all probability, more people lost their lives in hurricane disasters in this period than at any other time in the state's history. But because the commercial class did everything in its power to downplay and minimize such disasters, it is difficult to find reliable death and damage figures. This much we do know: Major hurricanes struck South Florida in 1926, 1928, and 1935. At the very minimum, the three storms together killed 2,487.[22] The 1928 hurricane (category 4)—the fifth most intense to strike the mainland United States—alone killed at least 1,836 people, most of them black migrant workers who had come to the Everglades a short time before for planting season. The 1935 storm (category 5) holds the record as the most intense hurricane ever to strike the country, the barometer falling to 26.35 inches. At least 408 people died, including 256 impoverished war veterans, sent down to the keys by the Federal Emergency Relief Administration to work on building a highway.

How much destruction the three disasters caused is very nearly impossible to calculate. The 1926 hurricane was the costliest by far, causing more than $1.4 billion in damage when converted into 1994 dollars.* While the other storms were far less destructive, the losses were high enough to turn heads in the insurance industry, prompting some companies to shy away from windstorm coverage altogether.[23]

DON'T CALL US

Apart from causing catastrophic damage and killing at least 115 people in Miami alone, the September 18, 1926, hurricane left more than 15,000 homeless. Yet it didn't take long—scarcely two weeks—before the *Miami Herald* declared that things were back to normal:

> The normalcy that has come is in the minds of the people. They are still just a little bit dazed, but they are sane—there is an absence of any of the hectic outbreaks that nearly always accompany disaster.... Miami has not yielded to the emotional temptations that come in such a time as has been experienced here. Mental normalcy means the most rapid rehabilitation of the city that is humanly possible.[24]

---

* According to a recent study by Roger Pielke Jr. of the National Center for Atmospheric Research and Christopher Landsea of the National Oceanic and Atmospheric Administration, were a storm of similar intensity to follow the same track and strike South Florida today it would cause $72 billion in destruction, making it the costliest storm ever, well over twice as expensive as Hurricane Andrew, which caused an estimated $33 billion in damage (1995 dollars).

The *Herald*—launched in 1910 by lawyer and businessman Frank Shutts to support the prodevelopment intentions of the railroad builder Henry Flagler (of whom more in a moment)—had a long record of downplaying events that shed a negative light on Miami. Indeed, when the Associated Press estimated that the city experienced $100 million in damage—as it turns out, a reasonable estimate—the *Herald's* managing editor, Olin Kennedy, directed his subordinates in the newsroom to divide the figure by 10. Ultimately, the paper printed the absurdly low number of $13 million, minimizing the disaster in the hopes that people would stay calm and, even better, forget it.[25] "By January 1," the paper predicted, "we will have ceased to think or talk of the hurricane. It will have written its story in the history of the city, but Miami will be looking forward, not backward."[26] Forward, that is, to a society where disaster was normal—part of the price for developing land in hazard-prone locales.

Some said that Florida had it coming. Gov. Austin Peay of Tennessee viewed the hurricane as a form of retribution meted out to Floridians who had lost "their hold on religion and law."[27] Letters received by the *Herald* also implicated divine forces, pointing to various sins—nightclubs, racetracks, Sabbath breaking—found in Miami. Hogwash, said the *Herald*: "The fact of the matter is that calamities come because people happen to be in the places where certain manifestations of nature take place."[28] In other words, the paper blamed nature, an interpretation of course that glossed over the human economic forces—the dredging and pumping of sand—responsible for placing all this property in the way of danger in the first place.

Much as their counterparts had in Charleston and San Francisco, the Florida business class sought to normalize the calamity by encouraging people not to panic or spread rumors. A proclamation issued by James Gilman, the acting mayor of Miami, and F.H. Wharton, the city manager, observed that despite the crisis, people "have maintained a calm feeling which is a manifestation of the same spirit upon which the civic development of our city is built."[29] Of course the declaration of martial law in Dade County undoubtedly forced serenity down people's throats. Maj. Robert Ward, the officer in command of Dade, urged citizens to "Get the truth in the newspapers: Do not believe the rumors heard throughout the city and spread by irresponsible persons."[30] At one point, the *Herald* published a front-page cartoon titled "The Wind That Does

The *Miami Herald's* effort at damage control (*Miami Herald*)

The Most Damage," satirizing all the many rumors circulating throughout the city. "Exaggeration—that is the latest foe to be conquered," editorialized the paper. Reports of thousands dead, of bodies littering the
streets, of massive starvation and epidemics, "all of this, as we know, was
greatly exaggerated," the paper continued. "Miami is not in ruins. Miami
will be her smiling self again within a short time."[31] In fairness, truth was
in very short supply. The *New York Times* carried a story reporting 175
bodies in temporary morgues in Miami soon after the storm, 60 more
than the official count at the time. In tallying the official numbers, however, government leaders admitted that "the names of negroes were not
always included."[32]

Yet when it came to physical destruction, even the *Herald* probably would have granted the hurricane's enormity. Eyewitness accounts printed in out-of-town newspapers such as the *New York Times* and *Washington Post* told of cars ripped to pieces, steel beams twisted into knots, ships lifted from the harbor and carried to shore, and thousands of homes and buildings damaged or destroyed.[33] To deal with the devastation and homelessness, the city of Miami, led by Mayor E.C. Romfh and Frank Shutts, issued a nationwide appeal for money.[34] Then they had second thoughts. By the end of September, Romfh had sent a telegram to the mayors of all major U.S. cities thanking them for their offers of assistance. "After inspection of the damaged region," he wrote, "I find the situation not as bad as reported. Will call on you if the need arises."[35] He didn't call. As a result of Romfh's actions, the mayor of Richmond, Virginia, threatened to stop a $10,000 check until he could figure out who was telling the truth, the "boosters or the relief workers."[36]

For his part, Gov. John Martin also downplayed the extent of the damage and even went so far as to refuse to call a special session of the legislature to appropriate relief funds.[37] But perhaps the most revealing example of the twisted logic employed by the minimizers came from the lips of Peter Knight, a prominent businessman, owner of the *Tampa Tribune*, and a major figure in Florida boosterism. Upset that many observers viewed the hurricane as a "Florida" disaster, Knight mused as follows:

> When San Francisco was visited by an earthquake, the press of the country did not talk about the California disaster. When Chicago had its great fire, the press of the country did not talk about the Illinois disaster. When Galveston was destroyed, no one spoke of the Texas disaster.... And now when but a small portion of Florida has been affected by the hurricane, the country refers to this as the great Florida disaster.[38]

The reasoning behind all this disaster nosology eluded Knight. It apparently never occurred to him that an event involving several hundred deaths and major damage throughout South Florida constituted a "Florida disaster." As he wrote, the Red Cross "in its zeal to raise additional funds" must be careful not to "do more damage permanently to Florida by the representations it makes throughout the press with reference to our conditions than will be offset by the moneys received."

Knight's contribution in this regard involved calling the $100-million damage estimate—again, an accurate figure—"simply absurd."[39]

Of course, such statements were hardly helpful to the Red Cross. For the first time in its history, it failed to solicit the requisite funds to relieve the suffering. As of mid-October, only $3 million of the $5 million in needed funds had been subscribed. The shortfall led John Barton Payne, the chairman of the Red Cross, to accuse state and city officials of placing real estate interests ahead of the needs of the storm's victims. "The officials of Florida from the Governor down and the real-estate operators have seriously handicapped the American Red Cross in its efforts to provide relief for those who suffered in the hurricane that swept southern Florida on September 18 by minimizing the loss," Payne told reporters, adding that "the poor people who suffered are regarded as of less consequence than the hotel and tourist business of Florida."[40] In a letter to Romfh, Payne charged that the city's attempt to downplay the damage had "practically destroyed the campaign to raise funds, and aside from this real estate people here and elsewhere have given out repeated statements to the effect that the damage was slight."[41]

Romfh defended his actions, explaining that he had no intention of "minimizing the seriousness of the situation." But as he also noted, "we in Miami, who depend upon the tourist trade for support to a large extent, could not afford to let the impression get abroad that the whole city was in ruins, when in reality, except for some of the outlying sections, the damage was not extensive."[42] Interestingly, Romfh himself was not one to minimize or deny the threat that hurricanes posed when it came to protecting his own business interests. Prior to the disaster, the First National Bank of Miami planned to open a new branch in Miami Beach. Romfh—the bank's president—had the building constructed three feet above street level to protect against hurricane-induced flood waters. "This caused intense amusement among more recent Florida arrivals," noted the *Saturday Evening Post*, "and the more facetious ones declared openly that Mr. Romfh's idea in forcing people to walk up three or four steps to get into his bank was to try to make them think they were in New York." But Romfh had the last laugh: The damage to the bank in the storm was minimal.[43]

Efforts to minimize the calamity and slow the flow of money amounted to a form of class warfare waged against South Florida's working poor. The further one went from Miami, the more evident this fact became, as we will see in a moment. In Homestead, located south of the city, the *Enterprise* explained that the stifling of aid caused Florida to bear increasing resemblance to, of all places, England—and under feudalism no less. "England in feudal times had food and comfort and pleasures for the wealthy, while outside the gates the peasants starved. Florida in this day can handle all the tourists that she did before—and then some, while another class of people are living from hand to mouth, dazed and bewildered at the change that has taken place."[44] In fact, the state simply typified the United States under early twentieth-century capitalism.

THE HAND OF PROVIDENCE

If the Red Cross's failure to meet its $5-million goal hurt anyone, it was the people of Moore Haven, a small town on the southwestern shore of Lake Okeechobee. The town took a vicious pounding as the 1926 hurricane tracked northwest from Miami. With the storm raging in the middle of the night, Moore Haven residents stumbled out of bed to reinforce the dike that protected them, breaking their backs to raise it two feet. It was a futile, not to mention a self-destructive, act. By morning, the dike began to break up; the all-night vigil to add to its height only served to increase the force of the water that eventually rushed forth. Exactly how many died in the flooding is unclear. The official count was 150. But a very knowledgeable local source said 300 died, in a town with a total population of just 1,200.[45]

Howard Sharp, for one, was not surprised. A self-described "tramp printer" who had visited earthquake-stricken San Francisco before venturing to Florida to take over as the editor of the *Everglades News*, Sharp predicted the 1926 disaster, what is surely one of the most accurate prophesies of doom on record. To fully understand his prediction, we must understand a little about the history and the stakes involved in settling this part of Florida. We begin with Napoleon Bonaparte Broward, who early in the twentieth century came up with the idea of reclaiming the Everglades. Under Broward's leadership, the state embarked on a more than two-decades-long drainage project. By 1929, close to $18 million had been spent to build over 400 miles of canals and levees with but one purpose in mind: to transform the rich muck lands into valuable farms. As railroads forged their way to the south shore of the lake, small

frontier towns—Pahokee, Belle Glade, Clewiston, and Moore Haven—
began to sprout.[46]

To say the least, it was not an entirely safe place to live, although it
may well have seemed hazard free to many at the time. During the teens,
as development near Lake Okeechobee proceeded apace, there were no
hurricanes or floods to speak of. Then, in 1922, prolonged, heavy rains
caused the lake to rise more than four feet. Clewiston and Moore Haven
were both flooded, compelling the construction of a muck dike along the
lake's southern shore. Two years later, another period of intense storm
activity raised the level of the lake, causing more flooding.[47] Finally, in
the summer of 1926, before the September hurricane, heavy rains again
raised the level of the lake, leading Sharp to beg state drainage officials to
take steps to lower the water. "The lake is truly at a level so high as to
make a perilous situation in the event of a storm."[48]

None of this seems to have concerned the Everglades drainage district
commission. Headed by some of the highest officials in the state, including Governor Martin and Attorney General J.B. Johnson, it took no
action to lower the water. And this even though as early as September 1,
the level of Lake Okeechobee exceeded 18 feet. The levees around the lake
were built to only 21 feet, and anyone even remotely familiar with the
area knew that a stiff wind could cause the lake to rise as much as 3 feet.[49]
The mathematics of fatality and destruction were painfully obvious. Yet
the drainage commissioners refused to act.

Apart from the wet weather, what kept the lake high was the sand
and muck that slowly filled in the canals, obstructing the flow of water
out of the lake. Of course, it took money to dredge and repair the caved-
in banks responsible for the siltation, money that had to come from more
taxes, which the big-time landowners, who bought up huge tracts around
the lake on speculation, were not inclined to pay. Instead, they sent their
lawyers to Tallahassee to bend the ears of the drainage commissioners, to
tell them, "Let the lake be damned, just no more taxes."[50] Meanwhile, the
poor, unrepresented workers and truck farmers who scrounged out a living around the bloated body of water found themselves squeezed to the
edge of extinction. Rarely are the man-made forces behind "natural" disaster so direct and obvious.

Which is precisely why state officials were so vehement in their
denial of responsibility for the death at Moore Haven. Instead, they
blamed nonhuman forces. Attorney General J.B. Johnson explained:

"The storm caused the loss and damage.... It is not humanly possible to guard against the unknown and against the forces of nature when loosed."[51] As Governor Martin explained, nobody

> can ever guarantee that the hand of Providence that sends winds and earthquakes and rains will have no effect, more or less, on this area, and the public is advised that now and for all times the Everglades area is subject to the same rules of Providence, and is under the same guiding hand that other lands of the earth are, namely, that of God Almighty.[52]

The fact that the state had both subsidized and encouraged settlement around Lake Okeechobee seemed not to cross the mind of either man.

When he was not busy blaming God for the calamity, Martin pointed his finger at nature. He publicly stated that a wind he clocked at 125 to 135 mph blew across Lake Okeechobee on that fateful September day. How he derived this figure is anyone's guess. No device existed for measuring wind speed on the lake during the storm.[53] From Martin's perspective, blame for the calamity rested with either God or nature, and both choices yielded the same result: an evasion of very real responsibility.

SURFACE DAMAGE

Two years later, in 1928, South Florida again found itself crawling out from beneath the wreckage caused by yet another tropical depression. On September 16, a powerful storm, with a barometric low of 27.43 inches—even lower than that recorded in 1926—swept ashore near Palm Beach, killing 1,836. Next to the notorious Galveston hurricane (at least 8,000 dead), it was the most deadly storm in twentieth-century U.S. history. Most of those who died were black migrant workers, virtually all of whom drowned in the towns along the southeastern shore of Lake Okeechobee, as the howling winds sent a wall of water crashing over the dikes.[54] Bodies were scattered everywhere. Governor Martin toured the devastated area and reported as follows:

> In six miles between Pahokee and Belle Glade I counted twenty-seven corpses in water or on the roadside but not taken from the water. Total dead on roadside and not buried and counted but not in plank coffins was one hundred and twenty-six. In six additional

miles over five hundred and thirty-seven bodies were already interred. Fifty-seven additional bodies were hauled out of this area today in trucks and tonight four truck loads of bodies were brought from adjoining areas by boat, loaded and sent to West Palm Beach for burial.[55]

Sightseers, brimming with morbid curiosity, filed into the region to see the mounds of swollen, rotting corpses firsthand. According to one report, "the visitor would stare for moments entranced, then invariably turn aside to vomit." Bodies were still being found more than a month after the disaster, when searching ceased for lack of funds.[56]

In an editorial after the calamity, the *Wall Street Journal*, seeking, in its own words, "to preserve a sense of proportion," educated its readers as follows: "Cyclone or hurricane damage is essentially surface damage. It has every element of the spectacular and it always looks several times as bad as it really is."[57] As the *Journal* put it in a subsequent editorial, "there is every need for sympathy but no need for hysteria."[58] At a distance of some 1,200 miles from the smell of rotting flesh—and with few *Journal* readers among the dead—it was perhaps easy to counsel restraint. Meanwhile, Peter Knight, no doubt eager to label this anything *but* the "Second Florida Disaster," remained similarly unruffled. He purportedly called the hurricane "trivial."[59]

Again, Howard Sharp seemed remarkably prescient, writing a week before the storm that those who advocated a high water level in Lake Okeechobee were taking "a terrible responsibility on themselves."[60] And again, a member of the Everglades drainage commission—this time Ernest Amos, the state comptroller—called the disaster an "act of God."[61] In truth, it was almost an impossible task for state authorities to manage the lake's water level to everyone's satisfaction. Commercial fishermen and those who used the lake for irrigation or transportation purposes wanted the water level high; farmers, especially those near the lakeshore, preferred it low to guard against flooding. Most of the large land speculators probably did not care either way as long as their taxes did not rise.

More than 20 years of Everglades drainage work and still the problem of managing the region's water remained. That fact spurred the state to seek the help of the federal government. In a 1929 congressional hearing, Florida Attorney General Fred Davis tried to explain the state's criminal neglect of flood control in the Lake Okeechobee area. Most residents of South Florida came from outside the state, he noted. Yet it was the people of northern Florida, settled much earlier than the flood-prone south, who dominated state politics. When it came right down to it, Davis told the committee, with most Everglades residents hailing from outside Florida, "it is mighty hard to get people in other parts of the State interested in whether they perish or not."[62] And perish they did.

BONUS SWIMMERS

Barometrically speaking, the most intense hurricane in the history of the United States occurred in 1935, the eye of the storm passing directly over the Florida Keys.[63] Once again, boosters were faced with spinning their way out of another "Florida" disaster. The state had experienced some devastating hurricanes in the last 10 years, the *Herald* lamented. "These tempests are thrust onto the front pages when they hit Florida, and thus they become Florida hurricanes." But it made little sense, the editorial continued, to try to convince people that the hurricanes did not origi-nate in Florida when, in fact, "the less said the better." As the paper con-cluded: "People forget rather quickly. It is wiser to let them do so."[64]

Perhaps not surprisingly, the approximately 400 people who died this time, the bulk of them war veterans sent down to the keys by the Federal Emergency Relief Administration (FERA) to fill in gaps in the highway between the mainland and Key West, exist in a kind of histori-cal no man's land. While the story of the unemployed veterans' 1932 march on Washington is well-known, few are aware of those who died during the 1935 Florida Keys storm. In their haste to rid Washington of these so-called bonus marchers, historian Gary Dean Best points out, the Roosevelt administration had not given "consideration to the conditions under which the veterans were being housed and cared for on the keys."[65] The housing provided the workers was in fact rather flimsy and certainly no match for the hurricane-force winds that accompanied the most intense storm in U.S. history. Nor were there adequate plans for evacu-ating the veterans, even though the risk of hurricane was widely known both in and outside of government.

When the hurricane hit, a storm surge estimated to range from 15 to 20 feet in height leveled the low-lying work camps. The blame game began almost immediately. Some indicted the U.S. Weather Bureau for failing to offer an accurate storm forecast.[66] But without more advanced weather-tracking technology—available only after the Second World War—precise forecasts were difficult. Others blamed the U.S. government for not having an adequate evacuation plan. By the time a train reached the keys it was too late to prevent it from being derailed by the high winds and water. For their part, Aubrey Williams of FERA and George Ijams of the Veterans Administration, the people charged with investigating the calamity, concluded that "responsibility for this disaster does not lie with any of the human factors concerned." As they saw it, in one of the most shameless uses of act-of-God rhetoric on record, "the catastrophe must be characterized as an act of God and was by its very nature beyond the power of man."[67]

Labeling the calamity an act of God outraged both veterans and religious groups. The Greater Miami Ministerial Association wrote President Franklin Roosevelt that they regarded the "statement that this catastrophe was an 'act of God' and that all was done that was humanly possible, as a deliberate attempt to whitewash known facts."[68] Rev. Frank Hamilton of the First Methodist Episcopal Church of Daytona Beach, Florida, told the president that he hoped he "as chief executive will not be satisfied to let this 'whitewash' stand, or to call it 'an act of God.'"[69] George Miller, chairman of the Veterans Union of Portland, Oregon, wrote Harry Hopkins of FERA that "this episode was not an 'Act of God'—but was an act of brutality and negligence perpetrated by responsible authorities."[70] There is no denying the administrative incompetence of those in charge of overseeing the welfare of the veterans. Nor is there any question that a more accurate weather forecast may well have helped to lessen or even avert the disaster. But in all the efforts to fix responsibility, it seems amazing that no one pointed the finger at the most fundamental reason for the tragedy: private-property-driven economic development and the accompanying progrowth attitude that opened the hurricane-prone keys to settlement in the first place. Pausing to briefly consider the history of this vulnerable area, one cannot help but conclude otherwise.

The railroad magnate Henry Flagler, more than anyone, helped clear the way for human occupation of the keys.[71] Having constructed a railroad from Jacksonville to Miami by the late nineteenth century, Flagler, in 1905, began working on what would be one of the more daring projects in railroad history. His ambition rested on extending his Florida East Coast Railway all the way from Homestead to Key West—a deep-water

Train sent to rescue veterans (Florida State Archives)

port—in the hopes of capitalizing on the anticipated increase in trade from the Panama Canal. The track would span roughly 130 miles, crossing 43 separate stretches of water or tidal marsh, one a full seven miles in length. It was an extraordinarily ambitious venture—some would say suicidal—and Flagler could find only one contractor willing even to discuss the project. The railroad took seven years to build, cost $20 million, and employed a stunning total of 40,000 workers. No fewer than 700 employees drowned in hurricanes while on the job.

That so many workers died is hardly surprising. Three hurricanes, in 1906, 1909, and 1910, struck the low-lying keys over the course of the project. In the fall of 1906, about 175 workers constructing the Long Key viaduct were living aboard a big wooden barge when it was smashed apart in a storm. According to the chief engineer, William Sanders, some of the men onboard became disconsolate and simply resigned themselves to death, drinking laudanum and going to sleep on tables, their pockets filled with the heaviest objects they could possibly muster. Flagler, meanwhile, remained undeterred by the disasters. After each storm, it is said, he telegraphed just two words to his men: "Keep going."[72]

Building in such a hurricane-prone locale was one thing. But the construction of the railroad track itself—completed in 1912—actually increased the danger. Erecting miles and miles of steel bridges and concrete viaducts proved expensive and, as it was, 17 miles of such structures were necessary. So to cut down on the need for any additional water-spanning technology, 20 miles of filled causeways were built, closing off the many gaps between the keys. Together, the newly built structures had the potential to work like a huge dike, obstructing the hurricane-induced ocean tide from flowing back out to sea and allowing the water to mount even higher than it normally would across the keys themselves. And that is precisely what happened during the 1935 hurricane. The practice of filling in between the islands had bottled up the water between the keys and the mainland, which increased the intensity of the damage and ultimately washed hundreds of people out to sea.[73]

Or so thought John Russell, the postmaster of Islamorada, Florida. Russell had lived his whole life—all 47 years of it—on the keys and was no stranger to hurricanes. He had been through the storms of 1894, 1906, 1909, 1910, and 1919. But the 1935 hurricane changed his life forever. He lost his mother, three brothers, all of his sisters, his wife, and three kids. Some 163 civilians living in the area, out of a total population of 270, died in the storm. "I don't believe that we would have had the loss of lives had there been openings in those creeks and channels instead of solid fills made by the railroad," Russell explained. "I have often said that some day the East Coast fills would drown the people if we ever got a very heavy storm which would back the water up." That day had arrived. "We being a community of poor people had no chance with a corporation," he said, referring to Flagler and his railroad so bent on filling in the gaps.[74]

The hurricane of 1935 changed more than just John Russell's life. It also helped to put a dent in South Florida's "see no hurricane" publicity strategy. The deaths of the veterans forced the region's hurricane problem into the national limelight. In testimony before Congress on providing compensation to the survivors of those killed in the storm, Congressman J. Mark Wilcox (D-Florida) admitted that 4,000 people had died in Florida hurricanes since 1925 alone, though he added, "it is not a good thing to talk about, it is a thing I hate to mention, because it is not good advertising." Asked if the state's "publicity facilities" had tried to minimize the hurricane threat, Wilcox responded, "There is no escaping the fact that back from 1925 to 1930, our people down there thought it was such bad advertising that we ought to deny the existence of them, just like the people of California think they have to deny the existence of earthquakes."[75] No Florida leader would ever have dreamed of making such a remark a decade before. By the mid-1930s, however, much as had happened with respect to earthquakes in California, it had become increasingly difficult to deny the hurricane problem. In part, calamity-torn areas were forced to admit disaster as they turned more to the federal government for relief funds, a common strategy after 1934, when Congress authorized the Reconstruction Finance Corporation to begin making disaster loans to rebuild public facilities.[76]

Openness about the region's hurricane risk, however, did not come easily to Florida boosters.[77] News reports filed after the 1935 disaster conveyed the impression to many outside Florida that Miami itself had been devastated by the storm, a worrisome prospect for the city's business leaders. "Miami and Florida object to having such storms designated as belonging to them," editorialized the *Herald*. "The business bodies of this community note that the tempest hurrying northward is often branded as the Florida hurricane, and think this is unfair, that something ought to be done about it, but what they know not. Neither do we."[78] In 1941, the prayers of Florida boosters were answered when George Stewart figured out a way to put a new spin on hurricanes. In his novel *Storm*, Stewart wrote about a young, up-and-coming meteorologist who names low-pressure systems after women.[79] During the war, the military used female names to identify typhoons in the Pacific.[80] And then in 1949 the U.S. Weather Bureau labeled a storm that smashed into South Florida, Bessie's Hurricane. The following year, the Weather Bureau formalized this practice of gendering hurricanes, solving, albeit unintentionally, Florida's adverse publicity dilemma.[81] There was nothing inevitable about this decision to feminize the hurricane problem. It was, instead, the product of the sexual politics of the postwar period, a time, as Elaine Tyler May has observed, that saw increasing concern about the destructive effects of female sexuality (especially as married women remained in the labor force after the war) and its effects on the family. What better way to demonstrate the evils of liberated womanhood than to give one of nature's most destructive and uncontrollable forces a female name?[82]

Transforming what had once been known across America as "Florida hurricanes" into female storms served to naturalize further the destructiveness of these calamities. That was something city boosters in hazard-prone areas—in their efforts to soft-pedal disaster—had been working for since the late nineteenth century. Women hurricanes were routinely described in the 1950s as wild, capricious, fickle, whimsical, and erratic, creating the sense that nature was literally out of control, when of course economic development, driven by private property, was as much if not more than nature to blame for disaster.

INTERLUDE

....

Body Counting

Charleston, San Francisco, and Florida are just three points on the compass of disaster. But what about all the other calamities that spanned the period between 1880 and 1930? What, by way of summation, can be said about them in the context of this study?

First and foremost, deaths skyrocketed in the half century after 1880, and, it seems safe to say, despite the absence of totally reliable data, that more people were killed in natural disasters during this time than at any other point in American history.[1] In this sense, the 1906 earthquake—the deadliest such seismic disaster in the United States—was just one of a number of extremely lethal calamities that define this period.

Before the Civil War it was rare for a natural event to lead to a major loss of life. The 1812 New Madrid earthquakes (M 7.6 and 7.9), for example, were estimated to be the most powerful earthquakes in terms of energy released ever to strike the conterminous United States and were felt over nearly one million square miles. Yet they killed very few people.[2] Similarly, an 1815 hurricane, the most powerful storm to hit New

England since 1635, led to just a handful of deaths, quite unlike a storm of comparable magnitude in 1938 (category 3) that killed 600.[3] Nor were the devastating effects of tornadoes truly felt until after the Civil War, when continued westward expansion and population growth put many Americans and their property at risk.[4]

The half century beginning in 1880, however, proved considerably more bloody. Hurricanes were the greatest killer, with victims typically drowning in the accompanying storm surge. The five most deadly storms ever to strike the nation occurred beginning in 1881, when a powerful hurricane (category unknown) pummeled the Southeast, killing some 700.[5] Two storms in 1893 (again, category unknown) together took somewhere between 3,000 and 4,000 lives and possibly many more. Hurricane number one battered the outlying areas of Charleston and Savannah, home to poor black plantation and factory hands, killing 1,000 and perhaps as many as 3,000.[6] Hurricane number two plowed through the low-lying areas of Louisiana; again the vast majority of the victims were poor blacks.[7] And while it is common knowledge that the most lethal storm in the nation's history (category 4) plundered Galveston, Texas, in 1900, killing 8,000 and maybe 10,000 to 12,000 (mostly white victims), the pair of 1893 calamities, not to mention Florida's 1928 hurricane—1,836 dead, virtually all of whom were black migrant workers—are scarcely remembered at all.[8] Apparently, race has had a filtering effect on the collective memory of disaster.

Floods, too, exacted a high price in human lives during the period in question. It is well-known that the infamous Johnstown flood of 1889 roared through the Conemaugh Valley and killed over 2,000, most of whom were working-class residents of this factory town.[9] But there was also disastrous flooding along the Mississippi in 1892, with destruction that John Wesley Powell, the director of the U.S. Geological Survey, likened "to that of some memorable wars," though the number of lives lost is unclear.[10] In 1903, floods in the Midwest caused 100 deaths.[11] In 1913, the *New York Times* estimated that as many as 2,000 had died in Ohio alone as the rivers of the Ohio Valley surged over their banks. "The deaths of northern whites," journalist John Barry has written, "sensitized the country in ways that deaths of black sharecroppers did not," referring to a little-known flood the year before (number killed unknown) along the predominantly black-inhabited lower Mississippi region. Meanwhile, the 1913 disaster inspired Congress to appropriate additional money for flood control.[12] There were also significant flood-related deaths in 1916 (118), 1921 (143), and 1922 (215).[13] The notorious 1927 Mississippi River flood killed, according to official Red Cross figures, 246 people, though the organization did warn Herbert Hoover confidentially that the number was "not necessarily reliable."[14] The actual number of deaths, Barry notes, "almost certainly ran far higher."[15] Surely a relationship exists between the low body count and the demographics of the Yazoo-Mississippi delta (where the flood did the bulk of its destruction), populated overwhelmingly by black farmers, who some may have deemed not worth counting.

Finally, tornadoes added thousands more bodies to the ledger of fatalities. Of the nation's ten deadliest outbreaks, five occurred between 1880 and 1930, including the storm with the single highest toll in U.S. history, the tri-state tornado swarm of 1925.[16] Causing the bulk of its destruction in Illinois, Missouri, and Indiana, the flurry of storms claimed 747 lives and annihilated some towns so thoroughly that the dead had to be laid out in windows in the hope that a surviving passerby—there often being no next of kin or neighbors left—might recognize the body.[17] As for the general trend in tornado-related deaths, one finds that the average number of deaths per year hovered at nearly 200 from the 1880s through the 1910s before spiking the following decade at a little over 300 and then declining steadily over the rest of the century.[18]

To some extent, population movements were behind all the fatality. Beginning in the late nineteenth century, more Americans began moving into disaster-prone urban areas. The Gulf Coast from Brownsville, Texas, to Pensacola, Florida, for example, a region of extreme hurricane risk, experienced a surge in population around the turn of the century. So did the seismically active West Coast, as well as those western and midwestern cities subject to moderate to extreme levels of tornado risk.[19] But urbanization and the increasing settlement of the continent is not all there is to the story. Partly, the high death tolls are a function of the inability to forecast and warn citizens of these events given the rudimentary state of weather observation technology. In addition, this was simply a period of relatively high seismic and extreme weather activity.

In the last two-thirds of the twentieth century, however, natural disasters have become far less life-threatening. As Table 1 shows, there has been a sharp decline in the average number of deaths, from 414 per year in the period between 1920 and 1954, to 264 per year between 1955 and

TABLE 1

Deaths in Natural Disasters, 1920–1989

Years	Hurricanes	Floods	Tornadoes	Earthquakes
1920–24	10	469	1,322	0
1925–29	2,114	579	1,847	13
1930–34	80	146	1,018	122
1935–39	1,026	783	926	4
1940–44	149	315	835	8
1945–49	67	304	951	8
1950–54	217	293	884	14
Total	3,663	2,889	7,783	169
% of All Deaths	25.26	19.92	53.66	1.17
Total Avg. Annual Deaths		414		
1955–59	663	498	525	34
1960–64	173	242	229	131
1965–69	412	512	705	3
1970–74	146	1,032	703	65
1975–79	88	774	284	0
1980–84	28	651	272	0
1985–89	101	709	250	71
Total	1,611	4,418	2,968	304
% of All Deaths	17.32	47.50	31.91	3.27
Total Avg. Annual Deaths		264		

SOURCES: Paul J. Hebert and Robert A. Case, *The Deadliest, Costliest, and Most Intense United States Hurricanes of This Century (And Other Frequently Requested Hurricane Facts*, NOAA Technical Memorandum NWS NHC-31 (Washington, D.C.: Department of Commerce, 1990), 20; table 3.7, "National Weather Service Estimates of Flood-Related Deaths in the United States, 1916–1989" in *Floodplain Management in the United States: An Assessment Report*, prepared for the Federal Interagency Floodplain Management Task Force by L.R. Johnston Associates (1992), 2: 3–17; Robert A. Wood, ed., *The Weather Almanac: A Reference Guide to Weather, Climate, and Related Issues in the United States and Its Key Cities*, 7th ed. (Detroit: Gale Research, 1996), 97; and Bruce Bolt, *Earthquakes*, rev. ed. (New York: W.H. Freeman, 1993), 277–282.

TABLE 2

Average Annual Death Toll, 1920-1989

Years	Hurricanes	Tornadoes
1920–29	212	317
1930–39	111	194
1940–49	22	179
1950–59	88	141
1960–69	59	93
1970–79	23	99
1980–89	13	52

SOURCE: Derived from Table 1.

1989. Deaths in hurricanes and tornadoes—responsible for three-fourths of the mortality in the earlier period—decreased dramatically in the second half of the century. Doubtless, the weather service's implementation of its watch and warning program in the 1950s accounts, in large part, for the decline. A closer look at this data, however (see Table 2), reveals that the downward trend in deaths from hurricanes and tornadoes actually began in the 1930s, suggesting that better warning services alone are not completely responsible for the shift. Significantly, as it turns out, nature has been somewhat sparing during the last two-thirds of the century. Between 1871 and 1930, for example, 102 hurricanes and tropical storms struck Florida; between 1931 and 1992 there were only 78.[20] With respect to tornadoes, a better warning system, explains twister historian Thomas Grazulis, combined with greater awareness of the problem and a trend toward rural depopulation since the Great Depression, go a long way toward explaining the decline in deaths. Grazulis also points to another important and little realized trend. Since the 1920s there has been a geographic shift in tornado activity out of the thickly settled Southeast and toward the Great Plains, where better visibility and lower population densities prevail.[21]

There is, however, one glaring exception to the general downward-sloping death curve. Flood deaths actually increased in the last half of this century, comprising close to 50 percent of natural disaster mortality. A rash of severe flash floods in the late 1960s and 1970s accounts for most of the loss of life. Although the history of flash flooding has yet to be written, it seems likely that race and class may have helped pull some people under the tide. For example, the worst flash flood in the last quarter century occurred in 1972 in Rapid City, South Dakota, and claimed 238 people, fully 14 percent of whom were Native Americans, even though they made up only 5 percent of the city's population. Next in line in terms of overall mortality was the Buffalo Creek disaster that same year in America's very own internal colony—Appalachia. Heavy rain so thoroughly soaked an earthen dam put up by a coal mining company that it gave way, causing water to barrel through Logan County, West Virginia, ultimately sending 139 to their graves. A company official called the event an "act of God" in an attempt to conceal the mining corporation's role in the tragedy.[22] In fact, flash floods march to a far more secular rhythm brought on by chain saws and bulldozers. Strip-mine the landscape into submission and the floods will not be far behind, a point readily apparent to residents in poor, coal mining regions like Harlan County, Kentucky (5 deaths from a flash flood in April 1977), Johnstown, Pennsylvania (76 dead in July 1977), Pike County, Kentucky (3 dead in 1979), Harrison County, West Virginia (2 dead in 1980), and Belmont County, Ohio (26 dead in 1990).[23]

Yet it's the twister that continues—even in a climate of declining overall natural-disaster mortality—to function as a death penalty for America's poor. In all, tornadoes claimed some 850 people in the United States between 1981 and 1997.* Mobile home residents, mainly the poor and elderly who are attracted to such housing because of its low cost, accounted for more than one-third of the fatalities.[24] And this even though the structures make up less than one-tenth of the nation's housing stock. Although the manufactured housing industry would have you believe that tornado disasters are the product of freak natural acts— equal opportunity killers roaring willy-nilly across the landscape—the figures suggest a less democratic, more class-based pattern. So how is it that living the American dream has become a capital offense? One need not be a conspiracy theorist to detect certain unnatural political and

---

* Preliminary figures from the National Weather Service indicate 127 tornado deaths in 1998, 64 of which occurred in mobile homes.

economic forces at work here, which we will turn to in a moment. There's a politics to wind, in other words, that goes a long way toward explaining why those least able to afford it often wind up on the receiving end of nature's fury.

But no one, whatever his or her social class, can afford to become complacent. The potential for massive death that would cut across class lines is still a very real threat as more people collect in coastal areas and active fault zones. In the seismically active areas of Los Angeles, San Francisco, Salt Lake City, St. Louis, and Memphis, a major earthquake would be catastrophic. One recent study estimated that a repeat of the 1906 San Francisco earthquake would kill between 2,000 and 6,000 people. The same study projected 2,000 to 5,000 deaths in the Los Angeles area should just an M 7.0 earthquake occur along the thickly settled Newport-Inglewood fault.[25]

Which brings us to New Orleans, a place *Weatherwise* magazine recently dubbed "the Death Valley of the Gulf Coast."[26] Surrounded by water—the Mississippi River to the south, Lake Borgne to the east, Lake Pontchartrain to the north, not to mention numerous bayous and smaller lakes—the city exists, for the most part, either at or below sea level. Complicating matters even further is the fact that waters off the city's coast tend to be shallow (compared to those near Miami), a situation conducive to storm surge, as water is slowly stacked up during a major storm.[*] Worse yet, there are few escape routes relative to other metropolitan areas.

If such is the geography of disaster in present-day bayou country, history, too, is hardly cause for optimism. In 1915, a category 4 hurricane (packing winds of 131 to 155 mph) narrowly missed the city. The storm killed 275, making it the ninth most deadly hurricane to strike the mainland United States in the twentieth century. A dreaded direct hit by a storm of comparable magnitude would likely turn New Orleans into a huge lake 20 feet deep, with mass death a very real possibility.[27] The moral of the story is clear: Think twice before assuming that high death tolls are a thing of the past.

---

[*] Waves that are five feet high out in the middle of the ocean increase in height as they move toward shallower water, where the wave energy has less distance to travel before touching the sea floor.

II

FEDERALIZING

....

RISK

FOUR

Building for
Apocalypse

After Florida's 1935 Great Labor Day Hurricane, a monument went up in Islamorada in memory of those who perished. What little inclination existed to remember the catastrophe itself, however, was drowned out by an act of collective amnesia as buildings rose around the memorial. Homes and commercial buildings flew up just a few feet above high tide. Construction proceeded with little concern for wind resistance. The wake-up call came in 1960, when Hurricane Donna—nearly identical to the earlier system in terms of its path and a close rival in intensity—slammed into the Florida Keys. Although few lives were lost, financially, Donna was Florida's most destructive hurricane to date. "In the 25 years between Keys disasters," wrote Stephen Trumbull of the *Miami Herald*, "the few feeble voices for restrictions have been shouted down by the builders of shoddy if sometimes showy houses—and the fillers of tidal mangrove swamps for sponge-like subdivisions barely above normal high tides."[1] As architect John Stetson explained in 1960, "A man seeking a haircut in a barber shop is better protected by the laws of this state than is an investor either buying a building or having one constructed."[2]

**Islamorada, Florida, after Hurricane Donna, 1960 (National Oceanic and
Atmospheric Administration, Department of Commerce)**

The lack of an adequate state building code was one thing. Yet the
U.S. government too must share the blame for helping to manufacture
hurricane vulnerability on the keys. After the 1935 storm, the federal government provided funds for individual homeowners to rebuild. To be
sure, much of that rebuilding was done with an eye toward mitigating
the hurricane danger, in contrast to the speculative construction of the
1950s. But that said, the government's willingness to provide such money
would turn out to be the start of a steady and monumental transfusion
of federal cash for subsidizing disaster vulnerability. Especially after the
Second World War, the insatiable desire of developers for prime oceanfront real estate received an enormous boost as a host of federal programs
helped pay for bridges, roads, and water and sewer systems, ultimately
placing more people and property in harm's way than ever before in the
history of the Sunshine State.

In the postwar period, the federal government emerged as a major
player in the political economy of risk. And it made its influence felt not

just in the realm of infrastructure, but in other areas as well. When erosion threatened Miami Beach in the 1970s, the U.S. Army Corps of
Engineers—notorious for its obsession with waging war on nature—
intervened to haul sand to the area, a technique that has provided
Floridians with a false sense of security, even if it has helped to mitigate
somewhat the effects of storm surge. Meanwhile, with the corps off
renourishing beaches—a practice that tended to help wealthy hotel operators—another arm of the government, the Department of Housing and
Urban Development (HUD), was setting standards for mobile homes—
long the sacrificial lambs of natural disaster—which only demonstrated
its scandalous disregard for the poor. The survival instinct reigns supreme
in big bureaucracies like HUD and the corps, and survival, in turn, generally depends on amassing increasing amounts of power. To achieve this
goal, such institutions have mainly served those in society who possess
money and political clout, and not, naturally, the poor and working class
in need of social reform. There were, in short, big winners and big losers
in the federalization of risk.

Before 1935, boosters tried to portray hurricane calamity as anything
but a "Florida" problem so as not to compromise the region's commercial
future—a rhetorical strategy designed to sever hazards from space.
Prospective tourists and bankers were given a hard sell in an effort to persuade them that the state was not an inherently dangerous place.
Significantly, the federal government's entry into disaster response accom-
plished this very same goal. Lured in part by the prospect of federal dollars, boosters accepted the disaster label and welcomed the U.S. government's efforts to underwrite life on the edge of obliteration. But one
consequence of government intervention has been the transformation of
natural disaster from a localized problem into a national one. The risk
bound up with living in a hurricane-prone area was now shared among all
the nation's taxpayers. And as risk and space diverged it became harder to
locate blame when calamity did strike and more difficult to discern that
the federalization of risk did not benefit everyone equally.

BULKING UP

The 1935 hurricane marked the last time that a massively deadly storm
crossed paths with Florida. Since then, hurricanes have killed many fewer
people, generally less—far less—than 100 per occurrence. Even the most

intense storms to hit the state in the last 60 years—the 1947 Pompano Beach Hurricane, Hurricane Donna in 1960, and Hurricane Andrew in 1992, all category 4 storms—each caused just 50 deaths or fewer. Otherwise, death figures from the other 28 hurricanes making landfall have numbered in the single digits.[3] But this trend toward fewer deaths might as easily reflect a decline in major storms as it does improvements in warning and evacuation plans. Sixty storms, rated as either hurricanes or tropical storms, made landfall in Florida during the period between 1911 and 1950; only 40 storms struck the state between 1951 and 1992.[4] If we have entered a period of increased risk for hurricane activity, as some scientists now believe, Florida could easily reemerge as a vast killing field reminiscent of the early part of the century.[5]

While the postwar trend in deaths from hurricanes has plunged, property damage has tended to spiral upward. Prior to 1960, there was only one storm in Florida with damage over $1 billion (in 1990 dollars): the 1926 hurricane. However, between 1960 and 1975 alone, four hurricanes topped the $1-billion mark: Donna (1960), Dora (1964), Betsy (1965), and Eloise (1975).[6] Simply put, untrammeled greed for waterfront and other low-lying property explains in large part why damage has soared.

Consider Miami Beach.[7] Further growth of tourism in the 1930s caused the hotel business to boom. Ten hotels went up in the city in 1935, followed by 38 more the next year. By 1940 there were well over 200 hotels, forming a concrete wall right on the edge of the ocean. Then, in the decade after the Second World War, came another surge in construction. By the mid-1960s oceanfront lots were virtually nonexistent, driving the price of such property up to $5,000 per front foot. Needless to say, all of this expensive real estate sprawled out along the sea was, given the patterns of Atlantic hurricanes, a prescription for self-inflicted catastrophe.

Meanwhile in Miami Beach, where the city council had long been cavalier about the risk of natural disaster, developers were given a huge boost in their efforts to capitalize on the coast. In an attempt to encourage further development of waterfront property, in 1948 the city council—with the prospect of a hotel boom on the horizon and the disastrous 1926 hurricane but a distant memory—extended the bulkhead line 75 feet seaward. The move was a boon to beachfront property owners because moving the line (which marked the easternmost extent of private property) further to sea made it possible for them to construct more bulkheads and gain additional land. The ordinances extending the bulkhead line amounted to a series of enclosure acts, reminiscent of what happened to common lands in Britain back in the eighteenth century. Extending the line allowed hotels to expand the size of their lots by enclosing the foreshore—the area of land between the ordinary high- and low-water mark, land vested in the state in trust for the public—thereby turning it into private property for the exclusive use of hotel guests.[8] This extraordinary giveaway of public land to benefit private interests did not go unchallenged. The state of Florida (among others) sued the city of Miami Beach to stop it from issuing building permits on the foreshore. In 1953 the state won its case, reasserting public ownership.[9] But the ruling was too late to stop the hotels' land grab. "Virtually every ocean front hotel had already run out bulkheads, lines of cabanas or groins which fenced in their beaches," one report noted.[10] Allowing the hotels to build right up to the water's edge not only closed off public access to the beaches—a point we will return to in a moment—but also led to a further increase in the amount of property likely to be blown away during a severe tropical depression.

Development in Miami Beach mimicked the trend in Dade County, also the site of sustained growth during the 1920s and late 1940s. Some of the new land developments, such as Coral Gables (1925), were laid out on the Atlantic coastal ridge, an elevated area protected from inundation. But other municipalities, such as Hialeah (1925) and Miami Springs (1926), sprouted in areas that were once little more than saw grass marsh. In the 1940s, Hialeah Gardens, Medley, and Pennsuco emerged—though they had hardly any population at the time—shotgun incorporations that created three new and very flood-prone municipalities west of the coastal ridge.[11]

The unrelenting pressure to subdivide land continued in the decade between 1955 and 1965, as developers discovered the extreme marketability of dredge-filled marsh, popular among prospective homeowners seeking easy water access for their boats. With waterfront development raging up and down the coast, investors purchased and filled bay-bottom land in places such as upper Biscayne Bay and New River Sound in Fort Lauderdale.[12] Symptomatic of this rush to develop every available square inch of waterfront was a project hatched in the early 1960s, the brainchild of William Anderson of the Atlantis

Development Corporation, which laid claim to four coral reefs several miles off the Miami coast. Reasoning that they had a great spot for a resort on their hands, he and his associates decided to make a nation out of the reefs and call it Atlantis, Isle of Gold. Gold was precisely what these promoters had in mind for the reefs, calculating that efficient development could produce 200,000 feet of spanking new ocean-front property, worth some $1 billion. Claiming that the reefs were really islands, Anderson and his company deemed them subject to colonization under international law. In 1963, they staked their claim and erected four buildings on the coral, only to find three of them later swept away in a hurricane.[13]

The destruction of three-fourths of his investment should perhaps have given Anderson pause. But the promoter of Atlantis remained undeterred. More remarkable still, it seems that he was not alone in his ambitions. In 1965, Louis Ray and Acme General Contractors began dredging and filling on the reefs in an effort to create more landmass. They, too, were nation builders with plans to call their new country Grand Capri Republic. Asked about the reefs, Ray explained that he never secured anyone's permission before taking possession of them. He just went there and built two sand-filled caissons and a house, making clear his willingness to defend his newfound land from intruders. As he explained, "don't get me wrong by saying if I am going to attack the Coast Guard or Navy, but I was going to have some semblance of a defense and I was going to build it and claim it and I occupied it, and I was going to defend it to the best of my ability and I was going to own it. Now, am I a nation? Me and four investors?"[14]

A nation, no; a trespasser, yes. At least according to Charles Silver, a Coral Gables entrepreneur who also claimed the reef, which he named— what else—Silver Island. "If they get on Triumph Reef, we'll blow them off," fulminated Silver.[15] But before Silver could do so, Hurricane Betsy (category 3) did the job, roaring through in 1965 and sending Ray's two caissons out to sea.[16]

Whatever colonial ambitions remained after Betsy were dashed by the U.S. government when it decided that it could not afford to let the reefs be turned into either silver or gold. It brought suit to assert its dominion over the disputed area. In the battle to determine who really owned the reefs, the courts found that because they were completely submerged by mean high water they could hardly be considered islands. The reefs were ruled part of the outer continental shelf and belonged instead to the federal government, thus quashing this bizarre speculative venture.[17] Neither Atlantis, Grand Capri, nor Silver Island would make it to the United Nations.

EMBRACING DISASTER

If the federal government stood in the way of what was surely one of the more far-fetched schemes for developing a waterfront, on the whole, it proved itself a zealous supporter of life by the sea. Beginning in the 1930s and accelerating after the Second World War, the U.S. government subsidized land use on barrier islands by helping to bear the costs of constructing causeways, bridges, and water supply systems, in addition to providing disaster relief. In response, land speculation and population growth reached new heights.

Public sector funds, for example, were used as early as the 1930s to help bolster growth and development on the Florida Keys.[18] During the Great Depression, Gov. David Sholtz declared a state of emergency in Key West, opening the way for funds from the Federal Emergency Relief Administration, which sought to turn the city into a resort town. Then, in 1935, the Great Labor Day Hurricane obliterated much of the Overseas Railroad and so demoralized the Florida East Coast Railway that the company sold its right of way and bridges to the state. The Monroe County Toll Commission, with help from the Works Progress Administration, set about building a highway from Florida City to Key West—opened in 1938—providing better access to the keys and smoothing the way for increased population growth and development. After the war, the highway was widened and repaved to accommodate modern automobile travel. Meanwhile, the navy and the Florida Keys Aqueduct Commission collaborated on a pipeline for shipping water in from the mainland. Opened in 1942, the pipeline offered residents a steady supply of water for the first time. (Formerly, water had to be trucked into the area.) The federal government also played a role in helping to provide reliable electric service through the Federal Rural Electrification Administration. The combination of adequate access, water supply, and electricity helped to boost economic and population growth. Not including Key West, almost 14,000 people lived on the tenuous keys as of the 1960s, nearly three times the population in 1950.

All were there courtesy of the federal government, which paved the way for life in the city, but, as it turns out, with only one road out of town for use in an emergency.

Growth in South Florida after the Second World War happened during a period of relative calm in terms of hurricane activity. Yet when disaster did finally strike in the 1960s, federal legislation was already in place for helping Floridians rebuild. A law enacted in 1950, for instance, allowed the president to authorize disaster relief for reconstructing public facilities without seeking congressional approval. The government's disaster safety net was woven even more tightly three years later, with legislation authorizing the Small Business Administration (SBA) to make low-interest loans available during disasters.[19] Florida reaped the rewards of government subsidy, beginning in 1960 with Hurricane Donna. President Dwight Eisenhower declared the keys a disaster area following the storm, opening the way for millions of dollars to pour in to rebuild bridges, highways, and water lines.[20] The SBA meanwhile offered homeowners and businesses low-interest loans. And federal funds continued to flow into the state after hurricanes Cleo (1964) and Betsy (1965). "It's getting to be an annual thing, like football season," quipped Douglas McAllister, who oversaw the distribution of SBA loans at the Miami regional office after Hurricane Betsy.[21] Betsy, in particular, opened the pocketbooks of U.S. taxpayers. After the hurricane, Congress passed legislation allowing the SBA for the first time to cancel as much as $1,800 on a loan, effectively transforming that amount into a grant.[22] By the 1960s, a regularized system of federal disaster relief had emerged.

But federal largesse depended on an open admission of disaster, a point that still gnawed at Florida's boosters. In an editorial written after Hurricane Cleo (category 2) in 1964, the *Miami Herald*, historically a major force behind increased land development and tourism, lamented having to accept the disaster label to secure federal funds. Such disaster talk, the paper objected, might give the rest of the nation the mistaken impression that "Florida's Visitorland has been clubbed to its knees.... 'Disaster' talk is inaccurate, and it is injurious."[23] Meanwhile, Clarke Ash, the associate editor of the rival *Miami Daily News*, whined that for the sake of government money, "a region which depends primarily on the tourist business must be publicized all over the nation as a major disaster area."[24] It was perhaps possible for South Florida's tourist-based business community to tolerate the disaster cachet in return for federal aid. But there were other strings attached to government funding that Miami Beach's hotel owners at least found hard to accept.

ZERO BEACH

To swim in the ocean at Miami Beach's Singapore Hotel in 1959, you needed a swimsuit and a towel. Not to mention a ladder. The beach had ceased to exist, and climbing remained the only way down to the water.[25] By the 1950s, Miami Beach had become a gigantic oxymoron. The protective dunes and sandy beach—once the defining features of the area— had been reclaimed by the sea. In part, erosion, caused especially by intense storms, was to blame. Erosion developed as early as the 1910s, and intensified after the 1926 hurricane, prompting the installation of groins to trap and save sand. So many groins had been built by the 1970s that the beach looked like "a military obstacle course."[26] But mainly, the siting of hotels and other structures too close to the water held the key to this vanishing act. The extension of the bulkhead line in 1948 allowed hotels such as the San Souci, Cadillac, and Saxony, among others, to build their cement pools and cabana decks right across the foreshore.[27] Even after the 1953 legal decision forbidding such construction across public land, hotel owners continued to build right up to the sea. For example, in 1956, investors constructed the foundation of the Americana Hotel in Bal Harbour just 15 feet from the ocean. A seawall raised in front only added to the problem. As waves smacked the wall, the water was driven down, scouring out more sand with each hit than it would have if the tide was simply allowed to spread over the shore.[28]

In 1964, the Army Corps of Engineers decided to lend Miami Beach a hand with its beach woes. It came up with a plan to dredge sand from the continental shelf and pump it onto the beach, widening the stretch from Haulover Beach 10 miles south to Government Cut. Apart from improving recreational prospects, the corps' plan also aimed to provide hurricane protection by restoring the original protective dune. By absorbing some of a hurricane tide's total energy, the dune promised to lessen the impact further inland. Yet because the dune would be designed to withstand only a 70-year storm, the corps required Dade County to issue a disclaimer, informing residents of Miami Beach that the project would not provide total protection against a hurricane tide similar to the one in 1926.[29] The total cost of the project was estimated at $30 million, and the federal government would pay a little less than half. The project had the support of many, including Miami Beach's mayor in the late 1960s, Jay Dermer. But implementation did not start until 1977, mostly because of the opposition of Miami Beach's hotel owners.

Why would the hotel owners wind up fighting a proposal designed to mitigate the effects of hurricanes and improve tourism? The answer has to do with money and land ownership. The corps promised to contribute substantially more money toward the project if the newly created beach became public property, a proviso that made the hotel owners apoplectic. By the 1950s, the hotels had transformed the bulk of the beach—7.5 miles out of a total of 10—into their own private domain. "Miami Beach was built and became the World's No. 1 resort because of private beach ownership," explained Edwin Dean, executive director of the Southern Florida Hotel and Motel Association, and a vigorous opponent of the corps' plan.[30] "We do not need beaches. We have plenty of beaches," Dean told reporters in 1969.[31] That was in fact untrue, as proved by the hotels' own beach resuscitation plan, broached in 1967, to repair the area's many storm-damaged groins.[32]

But as Mayor Dermer pointed out, the hotel owners' approach offered no hurricane protection. Moreover, any further delay in implementing the corps' project could have proved disastrous in the event of an intense storm.[33] Of course natural disaster was far from the minds of hotel owners such as Ben Novack of the Fontainebleau, though Novack did have nightmarish visions of the corps' renourishment plan causing blinding sandstorms, sending tourists diving for cover.[34] Instead, the hotel owners were mainly concerned with keeping their beaches firmly ensconced within a no-trespassing zone. "Anybody who spends big money for an oceanfront room is entitled to a private beach," Novack remarked.[35] The thought of nonregistered guests venturing along the shore, at the city's expense no less, drove him wild. "Do we want a Coney Island here?" Novack fumed. "Maybe he's Coney Island merchandise," Novack said of Dermer. But there was no way that Novack, who had spent millions to make the Fontainebleau into a private luxury resort, was going to tolerate public access to his premises. As he put it: "Beaches are necessary to the people of this city, and we have them. We don't have the type of beaches they have in some native areas where there is no advancement or progress. We're not that type of place where you can walk along the beach." About the last point, at least, he was right. As far as hurricane protection was concerned, the logic of the corps' plan eluded him. "Those sand dunes won't stand up under hurricanes...," Novack remarked, "the sand will go into hotel lobbies, and we'll have to shovel it out."[36]

Eventually, however, the hotel owners decided that beach nourishment, courtesy of the federal government, was not such a bad idea after all. Their about-face on the issue in part resulted from the outcome of a lawsuit that ruled beach restoration of no harm, monetary or otherwise, to the hotels. Their reversal also probably stemmed from the dawning realization that they alone could not possibly bear the costs of restoring the beach.[37] In any case, their decision to support the corps' plan certainly did not grow out of any deep-seated concern for protecting the city from hurricanes or opening its beaches to the public.

In the spring of 1977, dredges located a few miles off the coast fired up their pumps, and not long thereafter brown seawater began to spew ashore on Miami Beach, the official start of the Dade County Beach Erosion and Hurricane Protection project. The project lasted four years and cost a total of $68 million dollars. Over its course, the corps dredged up some 14 million cubic yards of sand—four times the amount of construction material used in the Egyptian pyramids. By most accounts the project is a success. Now storm-induced waves must first crash into the restored dune, helping to protect Miami from inundation. But whatever benefits may accrue protection-wise may be outweighed by an increase in hurricane apathy. Beach nourishment covers up storm damage and thus helps to disguise the true risks—such as the threat to life from storm surges—that result from living on the edge of disaster. In any case, even the corps itself admits that the new beach will not last forever. Erosion of a restored beach is estimated to occur 10 times more quickly than a natural shore.[38] And with the beach in constant need of replenishment, the offshore sand is going to run out. Indeed, by the early 1990s, officials in Dade and Broward counties, concerned about the dwindling sand supply, floated a plan to ship it in from the Bahamas. But there are obstacles to such a scheme. First, the cost of mining such sand would be extremely high. Second, there are federal import rules to consider. And third, since the Bahamian sand is mostly limestone, mixing it with the quartz-based Florida sand and subjecting it to the region's heavy rains could turn the material into cement.[39] Which, if nothing else, would solve the erosion problem.

LIVE FREE AND DIE

After Hurricane Betsy in 1965, South Florida lingered for almost three decades as if in a perpetual eye of a storm. In the Miami area, the lull in hurricane activity proved even more dramatic. The city had last been directly hit by a mega-storm in 1950, making for 42 years of quiescence before Hurricane Andrew (category 4) delivered its awful blow.[40] Perhaps not surprisingly, the peace and quiet paralleled the silent evisceration of South Florida's building code.[41] The code (passed in 1957 and applicable to all municipalities in Dade County) has long been considered one of the most stringent in the country, with rules requiring new buildings to withstand winds of up to 120 mph.[42] But starting in 1970, county officials slowly watered down the code under pressure from builders pushing for cheaper and quicker ways to construct (allowing, for example, the use of Masonite siding on exterior walls without any plywood backing).[43] The loosening of regulations, in turn, spurred builders on as they expanded further south and west into Dade County. "The sad thing is that in a hurricane-prone community like this, the builder should know he needs to be careful," said one Dade County civil engineer. "Alarm bells should go off in his head. People were just oblivious to things, as if they thought we never were going to have a hurricane in this area."[44] When Andrew did sweep through it exposed the incredible shoddiness. Plywood roof-sheathing required by code to have 33 nails had just 4.[45] "I have never seen so many people's lives put at risk and it's all for the sake of a few nails," remarked the civil engineer Peter Sparks. "If an engineer had done this, he'd be in jail now, or should be."[46]

Were one to hand down criminal indictments, first in line would be Dade County for its neglect of adequate mobile home wind-safety standards. Priced out of South Florida's conventional housing market, many low-income and elderly people found manufactured housing (as it is now called) very attractive because of its relative affordability.[47] Yet apparently affordable living has its price. The failure of mobile homes to stand up to high winds became amply clear in Florida after hurricanes Donna (1960), Betsy (1965), and Agnes (1972). Ninety percent of those showing up at shelters in Palm Beach County after Donna (category 4), for example, hailed from the metal boxes.[48] Although Betsy (a weaker category 3) spared densely settled Dade and Broward counties (saving its worst for Louisiana, where it killed 58 and injured more than 17,000), the storm still managed to generate an estimated 4,000 insurance claims in Florida. The number of claims, according to a postdisaster survey, would have doubled had Betsy's eye passed through the more densely settled parts of the state. Moreover, the same report noted that winds as low as 60 mph (that is, below hurricane force) seemed enough to mow down trailer parks, an observation that should have alarmed government officials at all levels.[49] Even after Hurricane Agnes (a category 1 storm that brought with it a series of tornadoes) killed nine Floridians, eight of whom had lived in mobile homes, Dade County officialdom remained unruffled. As late as 1973, the county still required only minimal frame ties, despite a recommendation from the Defense Department's Civil Preparedness Agency urging tie-downs that went directly over the top of the mobile home itself.[50]

The state of Florida itself was only slightly more concerned for the plight of the mobile homeless. In 1968, Florida, like a number of states, adopted a relatively weak construction standard put forward by the then industry-dominated American National Standards Institute.[51] Yet in 1971, at a time when more mobile homes were being sold in Florida (over 49,000) than in any other state in the nation, the government employed just three inspectors to oversee enforcement of the code.[52] Ultimately, public outrage over weak fire- and wind-safety standards drove Congress to intervene in 1974 with the passage of the Manufactured Home Construction and Safety Standards Act. But the national legislation, despite whatever good intentions existed, continued to spell bad news when the sky darkened and the winds began.

As a result of the act, the U.S. government found itself charged with establishing nationwide regulations aimed at reducing injuries and deaths. That may sound like a noble goal, but the act was flawed from the start. The legislation established a kind of lowest common denominator for standards, preempting state and local governments from upping safety and other design provisions on their own. Tremendous growth in mobile home shipments during the late 1960s and early 1970s had spurred many states to pass their own design, safety, and inspection criteria. Preemption, however, overturned these regulations and replaced them with a single set of federal rules, facilitating industry's ability to ship products across state lines. In some states, notably North Carolina, the federal standards trumped a code that was tougher and more safety conscious.[53]

In the deal cut by Congress in 1974, lawmakers granted federal pre-emption in return for language stipulating that the federal code would conform to the "highest standards of protection." But HUD's interpretation of the phrase "highest standards" left much to be desired. Consider the wind standards issue. After passage of the legislation, HUD divided the nation into two wind zones: standard and hurricane-resistive. In the latter, encompassing the coast along the Atlantic and the Gulf of Mexico, mobile homes had to be built to withstand wall and roof pressures of 25 and 15 pounds per square foot, respectively.[54] Suspecting that these wind load requirements might not be entirely safe, HUD asked the National Bureau of Standards to recommend improvements. Although the actual regulations do not mention wind speed, the bureau noted that the load rules translated into a design offering only 90-mph protection in hurricane-prone locales. The bureau issued a report, but HUD ignored it, turning a blind eye to the occupants of the region's ever more mobile homes.[55]

That especially impacted Florida, where mobile homes continued to remain wildly popular among the poor and elderly. During the 1980s, as refugees in need of housing poured into the region, the number of mobile homes in South Florida increased by one-third. By early the following decade, over 100,000 people in Dade, Broward, and Palm Beach counties lived in trailers, despite their extreme vulnerability to wind.[56] Meanwhile, prospective buyers were told over and over by sales representatives that mobile homes were "hurricane-resistive" and able to withstand 110-mph winds.[57]

Then, in 1989, two engineers, James McDonald and John Mehnert, uncovered even worse news. Mobile homes under the HUD design rules were barely able to withstand winds of 80 mph.[58] Although the mobile home industry insisted that their homes were essentially as safe as conventional housing, the boxes habitually flew through the air with the greatest of ease. "To be generous, you could say that HUD misunderstood the effects of wind on mobile homes," remarked Peter Sparks. "But it appears that they deliberately ignored the best evidence they were offered."[59]

When Hurricane Andrew came roaring through Florida in 1992, it left behind a huge pile of mobile home rubble, an unnatural disaster if ever there was one. In Dade County alone, the storm destroyed a stunning 97 percent of the more than 10,000 mobile homes there.[60] Four people died in them, versus a death toll of seven for conventional housing, even though almost 50 times as many people lived in the latter structures.[61] Still, the manufactured housing industry insisted on the relative safety of its product, blaming nature, not design flaws, for the mayhem. Edward Hussey of the Association for Regulatory Reform, an industry lobbying group, testified before Congress that Andrew was not by any stretch of the imagination your "typical hurricane." A press report claimed the suggestion that

> manufactured homes in particular were destroyed by only modest winds generated by Hurricane Andrew is dangerous and misleading.... There is not a single building code anywhere in the United States today which requires structures to be designed to withstand sustained winds ranging from 140 to 200 miles an hour.[62]

In fact, the most severely damaged part of Dade County experienced sustained winds of only 130 mph.[63] Measuring wind speed is not a simple matter, and a good deal of initial confusion arose over this issue immediately following the storm. Advocates for the manufactured housing industry seized on the confusion and conveniently chose extremely high wind-speed figures in order to help bolster their claim that nature, not design flaws, had caused the destruction. We now know that there may have been some peak gusts of 160 to 165 mph, but no good evidence supports wind gusts (much less sustained winds) of 200 mph.[64]

After Andrew, scientists at the National Institute of Standards and Technology (formerly the National Bureau of Standards) got out their computers and calculated what the HUD design standards translated into in terms of effective wind protection. It emerged that the 1989 study by McDonald and Mehnert was valid after all.[65] HUD was moved to action. In early 1994, the agency issued new wind standards. This time dividing the nation up into three wind zones, HUD required manufactured homes in zone 3—covering parts of South Florida and other hurricane-prone locales—to withstand a wind speed of 110 mph. Although clearly an improvement, the new wind rules are not retroactive, and thus it will be several decades at a minimum before the nation's mobile home stock catches up (somewhat) with what nature can dish out.[66]

Of course the new wind rules were hardly welcome news in industry circles. Frank Williams, the executive director of the Florida Manufactured Housing Association (FMHA) called the new standards an "overreaction" (to a disaster that obliterated nearly all of the mobile home stock in Dade County!) that threatened to price people out of the housing market.[67]

Outraged by HUD's decision to up the standards, the FMHA, in league with others in the industry, sued to stop the move. They argued that the new wind rules were simply another case of "governmental paternalism" designed to foreclose on people's ability to live cheaply and dangerously if they wished. To advance their cause, they drew an analogy between the wind regulations and auto safety standards for convertibles—criteria Chrysler challenged in court in 1972. But the federal appeals court handling the case rejected their argument. It distinguished the two situations, noting that in *Chrysler* the challenge was being made to legislation that affected only the occupants of the vehicle itself. The manufactured housing act, however, was designed to protect both mobile home dwellers and the general public from the threat of flying debris. As the court put it, "What the manufacturers propose would be the equivalent of allowing automobile purchasers to buy at a discount automobiles with unsafe brakes, a consumer choice option that would sacrifice the safety of innocent people."[68]

Although the industry lost its legal challenge, the tougher wind standards did give it a new selling angle. Said Robert Young, president of the FMHA, "Now, manufactured houses are a lot stronger. It makes them just that much better."[69]

In the 1990s, when Terry Trexler, president of Nobility Homes (also a party in the lawsuit), looked to sell his company on Wall Street, he liked to point out that the homes his company built cost less than the cars that stock analysts drove.[70] But the homes, while inexpensive (the average price today is $35,000 per unit), are not as durable or as safe as fixed-foundation structures. Is it any wonder why the industry has been forever dogged by an image problem? "People hear manufactured housing and they think of that trailer that used to sit out in the field behind Uncle Jed's shed," explained Jack Slater of the FMHA in 1992.[71] "We wish there was some way to make a major breakthrough [in rescuing the product's much maligned image]...," Frank Williams once said. "I know how the NAACP felt."[72]

In the world according to industry flacks, people *choose* to live in mobile homes. They are an affordable form of housing—no argument there—and some simply decide of their own free will to adopt this style

Dadeland Mobile Home Park after Hurricane Andrew, 1992 (National Oceanic and Atmospheric Administration, Department of Commerce)

of life. Explained Young in 1994, "Many people choose to live in manufactured homes because they get more for the money, enjoy the close-knit lifestyle of manufactured-home communities, or like the ease of upkeep and maintenance of this type of dwelling."[73] From this assumption about the volitional nature of mobile home life, it is but a small leap to the conclusion that increasing safety standards will exert an unwelcome influence by driving up prices and shutting people out of the housing market.

In the industry's view, the key question is, as the FMHA's Williams once put it, "How much safety can people afford?"[74] The real question, however, is how much safety can the industry tolerate and still have a price-earnings ratio that will satisfy Wall Street. Industry boosters see mobile home living as tied up with the live-free-or-die ideology at the heart of the American way. But in this case live free *and* die would be a more accurate slogan. The industry's concern with the affordability of mobile homes is simply a way of legitimating a particular political economy of risk, rationalizing, in the process, why the poor pay so dearly for wind-induced disasters.

Early in 1998, tornadoes swept through central Florida, killing 42 people, 34 of whom spent the last moments of their lives watching their mobile homes disintegrate around them. Advocates for the industry were quick to point out that if you want safety you have to pay for it. "Do you force people to pay twice as much for a home for a freak of nature like this?" asked the incorrigible Frank Williams.[75] Two things are important to notice here. First, as long as the homes are priced the way manufacturers want them to be priced, people are *choosing* to live in this form of housing. When the price is doubled, however, suddenly people are being *forced* to do something, coerced into buying a house that might actually afford them some protection during a windstorm. Second, explaining the very high proportion of mobile home deaths as the result of "freak" natural acts obscures the human forces at work here. In the industry's view, nature is portrayed as the predator. Yet it's hardly stretching things to say that the industry itself is doing the preying. In a rare moment of candor, a mobile home manufacturer admitted the obvious: "We don't build these homes to live in, we build them to sell."[76]

FIVE

························

Uncle Sam—
Floodplain
Recidivist

If Florida has invented itself as a hurricane disaster waiting to happen, Missouri—specifically St. Charles County, a place improbably located at the confluence of the Mississippi and Missouri rivers—has built its reputation on damage-defying acts of floodplain recidivism. Consider the Deerfield Village Mobile Home Park, which sits on a low-lying piece of land near where these two mighty rivers join. Home to hundreds of working-class Missourians who live there because it is the best housing they can afford, the park survived a major flood in 1986. In 1993, it was not so lucky. That summer, the Mississippi River surged over a nearby levee and submerged Deerfield Village. One could blame the disaster on the federally backed levee building and farming that deprived the watershed of its wetlands—nature's antidote to floods. Or point a finger at how federal flood insurance encouraged people to rebuild where they were destined to meet ruin again and again, a fact borne out by the county's record of repetitive claims, more than any other river community in the country. Miraculously, though, many persisted in seeing the flood as chiefly a natural event devoid of human intervention—a view destined to result in more rebuilding and more self-inflicted calamity. As Bob Bodley, a county building inspector, put it, "I don't hold anybody responsible."[1]

It would not have been difficult for people to acknowledge the role that development and other human economic forces played in bringing on the 1993 calamity. The emphasis nonetheless has tended to fall on the stunning combination of natural factors. What resonated, drew attention, and elicited remark was the set of immense natural forces—the high water, the heavy rains—that seemed to play a critical role in the flooding. At the peak of the disaster, *Newsweek* explained that the flood taught "everyone—even television anchormen—never to underestimate nature."[2] But what the flood should have taught people was never to sell short the power of the federal government (in league with various private interests) to create disaster. For what is truly striking about the destruction is the incredible series of events and policies that put people in harm's way. By 1993 St. Charles was a site begging for trouble.

To say that no one was responsible for the 1993 disaster is to deny the complex set of social and economic forces that has made St. Charles the flood capital of the nation. It is to deny the fact that risk does not simply exist, it is produced. Risk is manufactured, and who bears its burden is one of the most important questions we can ask in exploring the history of natural disaster. For all the people who consciously chose to make the floodplain their home, many more were forced to do so. The latter were compelled to live there by economic exigencies, by the simple fact that cheap, flood-prone land can be a magnet for the poverty stricken, who are forced to live in the shadow of disaster. It is one thing to develop a barrier island with hotels, to seek out a flood zone for the sake of profit. It is another to find oneself with little choice but to live in such a place. As disaster expert Lee Wilkins of the University of Missouri observed regarding the 1993 flood, "the riskiest thing to be was poor."[3]

DARK TRIANGLE

Almost half of the land in St. Charles County is located in the floodplain, an area with the potential to flood at some point, like it or not. Much of that land is in a triangle in between the Mississippi and Missouri rivers, which respectively form the northern and eastern boundaries of the county. Some people have been known to call this area the triangle of death, signifying the economic annihilation that has periodically resulted as floods have washed over the land. One look at this triangle and one sees immediately why it is so prone to inundation: It is as flat as a pool table. But that is not all. Runoff from almost one-fourth of the entire continental United States drains right onto that table, water from such far-flung places as Chicago and Montana and everywhere in between, all headed straight for St. Charles.

The first white people to settle permanently in this inauspicious locale came in the late eighteenth century, hailing mainly from Kentucky, Tennessee, and the Carolinas. But not until the 1830s did farm settlements sprout in any significant numbers in the region, as German immigrants forged into what they saw as attractive agricultural land. A local historian writing in the late nineteenth century called the area "the very Egypt of Missouri," referring to the rich, dark soil of the bottomland that was conducive to growing wheat and corn.[4]

The rich soil came courtesy of the rivers, which regularly flooded, bringing fresh silt and spreading it across the land. Water encroached routinely on the floodplain, but at times it took such complete command over the land that it inscribed itself on the historical record. Great floods along the Mississippi occurred in 1785 (the French called it *L'année des grandes eaux*), 1824, and especially 1844. In his *Centennial History of Missouri*, historian Walter Stevens described the 1844 flood as "greater than any that preceded it from the time the first record was made."[5] But only in the late nineteenth century did floods prove especially destructive to people and property in St. Charles. The *St. Charles Cosmos* described a spring flood in 1892 as "the most disastrous to property in the history of the two rivers. The country subject to overflow has built up rapidly the past few years."[6] More substantial levees rose in response, but 11 years later another great flood proved even more catastrophic, as a new railway embankment blocked the flow of water, causing it to be "mounted up with incredible speed."[7] The resulting inundation prompted John Buse, a St. Charles farmer, to observe that "thousands of acres of fertile fields of wheat in the last stages of ripening as well as a promising crop of growing corn were covered with from five to ten feet of water and were completely destroyed."[8]

The floods continued to devour the county in the twentieth century: 1908, 1909, 1915, 1917, 1922, and 1927. The 1927 flood submerged an estimated 10,000 acres, most of it in the notorious triangle of death area. Another 18,000 went under in 1935, as the Missouri River rose to more than 10 feet over flood stage.[9] By the 1940s, the county had developed a 100-year record of major floods.

None of this, however, seems to have harmed development. In fact, the number of buildings in the triangle increased, especially in the years after the Second World War, as groups other than farmers took up residence there. Two construction projects, in particular, helped to spur such growth. Project number one was the Alton Lock and Dam on the Mississippi River, completed in 1938. Part of a larger plan by the U.S. Army Corps of Engineers to improve navigation between Minneapolis and St. Louis, the Alton project was one of a series of dams designed to create pools of water that effectively gave the river a uniform channel nine feet deep.[10] But apart from its navigational value, the creation of slack water behind the Alton dam had another advantage as well. In 1944, Congress passed legislation allowing the corps to lease for recreational purposes land made prone to inundation by the new dams.[11] Upstream from the Alton project, three new subdivisions in St. Charles County went in along the river. The homes—originally built for weekend use but slowly converted to year-round housing—were generally built right on the ground and were thus excellent candidates for destruction by flood.[12]

Construction project number two began in 1956, when workers started up their bulldozers to make way for Interstate 70. The highway changed St. Charles forever. By tying the county ever more closely to the St. Louis metropolitan area, the road transformed St. Charles from a rural outpost into a bedroom community and created the need for much more housing. "That whole area started growing very, very fast after I-70 was built," remarked Bill Broderick, an engineer who helped to design the highway. "There were developers breathing down our necks all the time after that."[13]

With the county on the cusp of a major building boom, St. Charles residents voted in 1957 to put zoning into effect.[14] Meanwhile, developers rushed to get new subdivisions recorded before regulations on minimum lot sizes and septic tanks made it impossible to maximize profits. Much of the newly platted land existed in the floodplain, including numerous subdivisions as well as three sizable mobile home parks—Bedford, Princess Jodi, and Missouri Mobile. The vast majority of the new housing was built directly on or very close to the ground, and all of it was destined to take on water.[15]

ANKLE DEEP

By the 1960s there were two main groups living in the triangle: the farmers, who were reasonably well-off, and the poor, who lived in either mobile home parks or subdivisions along the Mississippi. The two groups had at least one thing in common: floods for both had become a way of life. Otherwise, their views about how life should proceed on the floodplain could not have differed more. In the end, their differences came down to the question of how the federal government should intervene to rescue the area from the ravages of nature.

A farmer appreciates a good flood every so often. Floods are nature's way of restoring fertility to the land, and fertility is of course the essence of successful farming. The problem with the triangle, however, was that the floods just kept coming. A flood every 20 years is something a farmer can perhaps live with. But in the 1940s, the river gauge on the Missouri at St. Charles exceeded 30 feet more than 100 times.* Nineteen fifty-one was another bad year, including a crest of 37.3 feet, the highest reading on the Missouri River gauge since official records began, in 1918. Things settled down a bit during the remainder of the 1950s, but in the 1960s the inundations resumed, causing the county's farmers to band together to do something about the problem.[16]

In 1967, the farmers established the North County Levee District, an organization with the power to tax for the purpose of keeping the thieving waters of their two rivers at bay. The district's first project involved raising money to build a levee 8.5 miles long. But this was only a stop-gap measure. To truly protect the 37,000 acres of rich farmland (and the mobile home parks and subdivisions by default), it was necessary to build something far more substantial, something the levee district could not afford to construct on its own. So the district waited to see if it could induce the federal government to become involved in its levee woes. The proposed L-15 levee—for left bank levee, 15 miles from the confluence of the two rivers—was slated to go 25 miles down the Missouri to where it reaches the Mississippi and then another 19 miles upstream as far as Portage de Sioux. If all went right, the Army Corps of Engineers—which until recently had rarely met a water control device it didn't like—would foot the bill.

---

* In actuality, flood stage at St. Charles is 25 feet, but only when the water reaches 30 feet in depth do things truly become disastrous.

Things did not go right. Opposition to the L-15 project emerged right from the start. Retired Col. Edwin Decker, who had once worked in the corps' St. Louis office, could not have been more blunt: "You must be out of your cotton-pickin' minds to even consider building a levee at the confluence of these two rivers," he told the corps. Building a levee and taking close to 40,000 acres of land out of the floodplain, he pointed out, would simply force the water into taking another route. That path would head straight through the Illinois communities of Alton and Grafton across the river—towns naturally opposed to the project. "Nature developed the floodplains as a safety device," explained Decker, "and now you want to take that away."[17] An Illinois-based environmental group named Pride argued that the levee would eventually lead to the industrialization of north St. Charles, a point the farmers who supported the project vehemently denied. Their intention, they said repeatedly, was to get the government to build a levee to protect agriculture, not to profit by selling out to industry.[18]

For the moment the corps took no action on the levee plan. Then, in 1973, a spring flood barreled through the county. The St. Charles river gauge registered readings in excess of 30 feet on 50 different days during the year, an all-time high. The flood drove 5,000 people from their homes, inundated almost one-third of the county, and destroyed the levees built by the North County district. Again, the farmers cried out for the L-15 project. Again, the environmentalists and Illinois communities opposed the levee.[19] Again, the corps sat back and did nothing.

If the 1973 flood demonstrated anything, it was that levees, far from offering protection, had actually increased the damage caused by the two great rivers. Or so concluded Luna Leopold, one of the most respected hydrologists in the country and the son of conservationist Aldo Leopold, whose *Sand County Almanac* remains the bible of many environmentalists. "I could never speak as eloquently tonight as your river is speaking to you now," Leopold told an audience gathered in St. Louis during the height of the 1973 flood.[20] The striking thing about this particular flood is that, compared with other major floods, it was relatively small in volume. Yet it resulted in some of the highest depths along the Mississippi ever. At St. Louis, the 1973 flood registered a depth of 43.23 feet and a volume of 849,000 cubic feet per second (cfs). That was, to be sure, a lot of water to accommodate. And if we search the historical record we find a flood in 1908 with nearly the identical volume, 850,000 cfs. But there is one critical difference: The crest of the 1973 flood was over eight feet higher than the crest in 1908.[21] According to Leopold, the high crest resulted from the building of levees and floodwalls, which held the water in and caused it to rise. If there was ever a time not to build a levee, this was certainly it.

The federal government refused to fund such a structure. But the government did become involved financially in helping rid St. Charles County of floods, only not in a way that pleased the county's farmers. In 1968, after decades of approaching the flood problem from a structural perspective—building levees, dams, and other water control structures and paying dearly for them—Congress decided to try something new: It passed the National Flood Insurance Act. The legislation provided federally subsidized flood insurance to residents in areas prone to inundation as long as the communities themselves adopted laws regulating land use on the floodplains.[22] The program thrived in St. Charles County, especially after the 1973 flood raged through the area. Policies written in the county increased from a few hundred in 1973 to over 2,000 by early 1975.[23] Although it was popular among the county's poor, the farmers mostly eschewed the insurance because it compromised the sanctity of that most cherished of American institutions—their private property.

The federal flood insurance program worked as follows: During the emergency phase of the program, a community instituted regulations to protect the floodplain from unwise land use, and its residents were rewarded with the heavily subsidized insurance. St. Charles entered this phase of the program in 1969. But a community could not stay in the emergency phase forever. The law required it to move eventually into the program's more stringent regular phase, something St. Charles did in 1978.[24] In this phase, the insurance premiums increased, especially for the more flood-prone areas. To calculate these rates, the federal government sent the Army Corps of Engineers in the mid-1970s into St. Charles (and other communities involved in the program) to draw maps depicting exactly what areas would be damaged during a 100-year flood. There was no surprise here. Roughly one-third of St. Charles County was at risk of going under in such an event. According to the federal guidelines, new structures in the regulatory floodway would need to be built so that the lowest level was one foot above the point where a 100-year flood would reach. The notion of a "100-year" flood makes you think it's something only your grandchildren are going to have to worry about. In fact, the expression is simply a shorthand way of saying a flood with a 1 percent chance of occurring in any given year. A flood this year does not mean immunity for the rest of the century; the same 1 percent risk exists year in and year out. This misleading formulation has had consequences on flood insurance sold, not to mention flood control projects, not just in St. Charles but throughout the nation, as the American public was sold a technocratic bill of goods with this optimistic phrasing.[25]

Meanwhile, for the towns of Portage de Sioux and West Alton— located near the tip of triangle, where the two rivers came together—the federal flood insurance program amounted to a death sentence. The entire city of Portage was built between 2 and 18 feet below the 100-year-flood level, and thus compliance with the program threatened to regulate the town out of existence.[26]

Given the stakes involved in flood insurance, could local officials in St. Charles really be trusted to enforce the stringent regulations? In 1982, FEMA decided to find out. Investigation showed that the county's board of adjustment and appeals had granted variances in half the cases where owners sought to build below the 100-year-flood level. So many variances were granted that FEMA threatened to drop the county from the program altogether. The high percentage of variances was no mystery to anyone who knew Herbert Meyer, the board's chairman. A farmer with land in the floodplain, Meyer himself did not carry any flood insurance. "I don't believe in the flood insurance program at all," he said. This from a man charged with implementing the law.[27]

As it turned out, in granting the variances, the board completely ignored the corps' survey of what constituted land at risk of a 100-year flood. Instead, Meyer and his colleagues on the board simply asked applicants if their land had flooded in 1973. If it didn't go under in 1973 that was good enough for them, and the variance was granted. There was only one problem: The 1973 flood was, even under the rosiest of scenarios, only a 50-year event. In other words, Meyer and his board had taken it upon themselves to lessen the standard of protection, effectively cutting it in half. In his defense, Meyer pointed out that the 100-year level employed by the corps had been chosen arbitrarily, which was true enough. Nineteen seventy-three remained the worst flood he could remember, and thus as good a standard as any for deciding where not to build, though it is worth pointing out that the crest in 1973 (36.2 feet) was a foot lower than the record one of 1951.[28]

Apart from Meyer, many local farmers groused about the flood insurance program. And for good reason. When the county signed off on the Corps' regulatory floodway, the prospects for the L-15 project went straight downriver. Complying with the flood insurance program's regulations increased the cost of building such a levee. Whenever the corps returned to study the project, it got to the end of the ledger and found that the costs inevitably outweighed the benefits. Flood insurance had cost the farmers their dream levee, at least for the moment.[29]

It did, however, help to underwrite life among the disadvantaged, who invariably wound up slogging it out in the floodplain. "You've got to live with what you've got," said one mobile home resident after a 1978 flood. "When you don't have the money, where else can you go?"[30] The federal insurance program subsidized the poor in a place where real estate capitalism had forced them to live—on the margins, in the cheapest, riskiest, and most flood-prone land. One of the worst spots in the entire county was just off Hawning Road, close to the Missouri, an area so prone to flooding that the corps ruled it off-limits to all new building. Predictably, this is where St. Charles's poor lived, in three mobile home parks platted back in the 1950s. For many at Missouri Mobile, the most wretched of the three developments, flood insurance helped to sustain what little material life they had.*

Much the same situation prevailed for those who lived along the Mississippi River as well. Holiday Harbor is one of the many subdivisions located on the Mississippi where flood insurance had allowed inundation to become a way of life. In the spring of 1982, when the Mississippi was running high, Bill Rainey would get up each morning and put on a pair of waders. Then he would carry his wife and two kids across their family room, out the sliding glass door, and into a boat and off to work. It was a tough commute. Eventually the water would recede, leaving behind a few inches of mud on the carpet. Apart from the carpet, Rainey and his wife expected to lose some furniture and a washer and dryer. They also expected the government to pay for the loss because they

---

* The residents of the Bedford and Princess Jodi parks rented lots and pads from the companies that owned the parks, and because they leased the land they were not entitled to federal flood insurance. At Missouri Mobile, residents owned both their homes and the lots.

had flood insurance. "Everybody's got it, if they've got any sense," Rainey remarked.[31] In 1982, over 1,700 flood insurance policies, valued at $43.5 million, were in force in St. Charles County, more than any other community in the state save Kansas City.[32]

FORGOTTEN COMMUNITIES

By 1986, St. Charles County had amassed an extraordinary flood history: 28 floods in 35 years. Getting your ankles wet had become the norm. Still, the 1986 flood marked out new terrain in the county's record book. The crest of 37.5 feet exceeded the 1951 record.[33] Nine hundred and thirteen mobile homes—17 percent of all those in the county—were either damaged or destroyed in the disaster.[34] The water in the Missouri rose so high that it temporarily changed the very identity of the river itself. Forced to enter an overflow channel, the water ultimately emptied out into the Mississippi River west of Portage des Sioux, 18 miles upstream from where the river normally disembarks.[35]

The magnitude of the 1986 flood compelled FEMA to take a hard look at the situation in St. Charles. Amidst all the high water and destruction, FEMA, realizing to what extent the county had mortgaged its future, took two moves to stave off future flood disasters. First, it sought to lessen the prospects for building the L-15 project by threatening to institute more floodplain rules.[36] Second, it tried, though not particularly hard, to regulate the siting of mobile homes more stringently. Nothing better demonstrates the man-made dimension of flooding in St. Charles than FEMA's approach to mobile home regulation. At least some of the massive destruction that occurred in 1993 could have been avoided had FEMA not been so ineffective.

After the 1986 disaster, FEMA initially urged St. Charles County and other flood-prone communities throughout the nation to elevate mobile homes out of danger. Prior to this time, existing mobile home parks were the only form of housing that FEMA allowed to be rebuilt below the 100-year base flood level and remain in the flood insurance program. This so-called grandfathering rule was what FEMA sought to end. Under the new rule proposed by the agency, should a flood or other disaster damage a mobile home by 50 percent or more, the structure could not be repaired or replaced and remain covered by federal flood insurance unless it was raised one foot over the 100-year-flood level.[37] Compliance with the rule would have required mobile homes in Bedford Village and Missouri Mobile to be elevated seven to nine feet.

Opposition to the ruling emerged from those with the most to lose, the big-time landlords who operated the mobile home parks. A lawyer representing Bedford Village said that were St. Charles to agree to the rule, it would amount to an unconstitutional taking, and litigation would probably result. In Missouri Mobile, the rule caused Glen Smith, a landlord who had managed to amass 32 lots, to bristle. "The river did a pretty good job, it took 50 percent of us," he said. "I don't need the county to take the rest."[38]

Some residents of the park, however, took an even more radical view than FEMA. They argued that the mobile home parks should be moved out of the floodplain entirely. Steven Shear of Bedford Village said the park "should be closed down." Let the county relax its zoning restrictions so that mobile home parks could be located in less flood-prone areas, he urged. Kathleen Hayes, another resident of the Bedford park, agreed. "I believe, like Steven, that they shouldn't allow mobile homes in high risk areas," she told the St. Charles County Planning and Zoning Commission, charged with deciding whether the county should adopt the new FEMA requirement.[39] The mobile homes were in hazardous areas because market forces had dictated that they be placed there. Ultimately, what the residents were advocating amounted to nothing less than an end to the prevailing political economy of risk created by real estate interests.

Eventually, the planning commission recommended halfheartedly that the county adopt the new regulation. "I'm making this motion with a great deal of reluctance, it's like a gun to our heads," said commission member James Beardsley.[40] Of course, *not* adopting the new rule amounted to pointing a gun at the head of every mobile home resident on the floodplain. This situation would prevail soon enough. The National Manufactured Housing Federation, representing the operators of mobile home parks and some of the major retailers, pressured Congress into suspending the regulation. Charging that the new rule would cost the manufactured housing industry some $500 million, the federation urged FEMA to study the economic impact of the regulation before proceeding.[41] FEMA, swayed by such hyperbole, backed off. It suspended the rule in June 1987, and by 1990 had instituted a much less stringent standard. New mobile homes sited in existing mobile home parks would be required to be raised three feet off the ground. For the mobile home parks near the Missouri River in St. Charles, a mere three feet hardly elevated the structures above the 100-year-flood level. Worse yet, because the new rule required mobile homes to be secured more permanently on reinforced piers, it threatened to make removal of the structures during floods even more costly and difficult.[42] By the time the 1993 flood came sloshing in, the residents of the parks located near the Missouri were sitting ducks.

Meanwhile, with FEMA caving in to the demands of the mobile home industry, saturated life in the parks continued. Paul Saunders of the state health department visited Missouri Mobile, with its 250 trailers in 1988, and found three inches of standing water in the streets. The park had no storm water drainage system because it was built before zoning required it. "The area is generally uninhabitable for many periods of time during the year because it is in the floodplain," said Saunders. "Any time it rains for a number of hours or the river is high, this area is going to have problems." Residents complained of as much as eight to ten inches of water in the roads. "We feel like we are a forgotten community," said Vicki Travis, a park resident and mother of three. "Our children have to walk through this water to catch the school bus. You can't walk to a neighbor's house. The cars get stuck. They don't deliver the mail some days."[43] It was third world living, and it was hard to imagine things getting any worse. But they did.

CARPET RENEWAL

As we have seen, even before the 1993 flood, St. Charles County had more repetitive losses than any other river community in the United States. In all, 899 structures had filed two or more claims under the National Flood Insurance Program since 1978.[44] "There are people in St. Charles County who collect insurance every time two guys upstream spill their beers in the river," said Ron McCabe of FEMA.[45] Some, including Tom Szilasi, St. Charles's building commissioner at the time of the 1993 flood, had grown tired of seeing the federal government continue to subsidize such madness. "Taxpayers have bought some of these people refrigerators and chain saws 10 times over," said Szilasi.[46] FEMA's idea of relief, he concluded, was "to stand on the back of a pickup truck and throw away sacks of $100 bills."[47]

No group felt more outraged by federal flood insurance than the farmers of St. Charles County. Over and over they pointed out that they themselves were not the main beneficiaries of the program. Instead, the money flowed to those who lived in cabins and mobile homes near the rivers. Yet the farmers' own property rights were sacrificed through the floodplain regulations required to stay in the program. "It seems we existed in this county up to 13 years ago without the federal agency, without the insurance, why can't we exist now without them?" wondered Ray Machens, secretary of the North County Levee District and a member of a long-time farming family, in the early 1990s.[48] Steve Ehlmann, once the lawyer for the levee district, took a similar view. When a flood occurs in St. Charles County, he explained, there are two groups of people, the farmers and the rip-off artists. As he put it:

> There's the one group who...the women and children go to church and pray that the rain stops and the men go out and start sandbagging. And then there's the other group, who the women and children go to church and pray that the rain doesn't stop and the men go to the bar and have a few drinks and talk about how they're going to spend their flood money.[49]

The pattern of repeat claims was making a mockery of the flood insurance program, and FEMA decided to do something about it. It picked the perfect time to make its move. With the county feeling the early effects of the 1993 spring flood, FEMA threatened to suspend it from the program if it didn't limit floodplain development.[50] In retrospect, the county might have considered thanking FEMA for at least trying to save it from itself.

The 1993 floods put 30 percent, or 128,000 acres, of St. Charles County under water. Some buildings remained afloat from April all the way through October, as the river gauge on the Missouri registered 60 readings over 30 feet. President Bill Clinton declared the county a disaster area no less than three times that year. More than 4,300 structures were either damaged or destroyed, and roughly 2,000 families were made homeless.[51] Worst of all, the disaster was completely predictable. Admittedly, the Missouri River at St. Charles crested at an incredible 40 feet, just a half foot below the 500-year-flood level.[52] But the 500-year-flood designation exists because a disaster of that magnitude is within the realm of possibility. Although such a massive deluge has only a two-

**Sandbagging in St. Charles, 1993 (Federal Emergency Management
Agency, Andrea Booher, photographer)**

tenths of 1 percent chance of occurring in any given year, we know it will
happen. We just don't know when.

Earl Hampton, a retired ironworker with a cabin in one of the subdivisions along the Mississippi, believed differently, however. "This
here's an act of God," he said of the 1993 flood and the destruction it
caused.[53] Hampton rented a lot in Sherwood Harbor and had been
there since 1956. He leased the land upstream of the Alton dam from
the Army Corps of Engineers, though technically it was for weekend,
not year-round use. But the corps had been ignoring that restriction for
a long time. The rent was cheap, just a few hundred dollars per month
as late as the early 1990s. According to Ron McCabe of FEMA, the
clubhouses, as they were called, were good places to "smoke a cigar,
drink a beer, and fish out the window."[54] They were also good places for
collecting flood insurance. Hampton had filed a claim in 1986, and he
did so again in 1993. As far as complying with any regulations and ele-
vating his cabin out of the floodplain, Hampton just laughed. "Hell,

you'd need an elevator to get in there," he quipped. "It could hurt if you fell off the porch."[55]

It is people like Hampton that Ehlmann and the farmers have in mind when they talk about the evils of the flood insurance program. Undoubtedly such abuses exist. Until recently, one heard stories about the "neighborhood washing machine."[56] In theory, people would move a beat-up old washer from house to house in advance of the adjuster, collecting insurance money repeatedly. Eventually FEMA started taking down serial numbers to stave off any further fraud. It would be a mistake, however, to see these worst-case scenarios as the norm. Most of those with federal flood insurance were not trying to rip off the system. In fact, if anything, the system was more likely ripping off the people. Recall that the entire rationale for the federal flood insurance program centered on the offer of subsidized insurance in return for more stringent land-use regulation at the state and local levels. But instead of holding communities to the bargain, the federal government has over the years simply weakened the standard to the point that by 1994 the Clinton administration simply required that "positive attitudes" with respect to floodplain management be "encouraged."[57] Paying for a few extra washers is one thing, but what the federal government has perpetuated on the American people here is the equivalent of passing out the keys to the treasury.

Before indicting the welfare queens of the floodplain, it's worth pointing out that in all probability many of these people were hardly there because they wanted to be. Instead, what drew them was the fact that housing is cheaper on hazard-prone land. The 1993 flood drove 4,500 residents out of their homes in six mobile home parks.[58] Candy Edwards and her two children were sent fleeing from the squalid Missouri Mobile park and found temporary quarters with a friend. In October, Edwards headed a drive to allow residents of the mobile home park to return. At the same time, she wanted nothing more than to get out of the floodplain permanently. "I'm not stupid; I know I have to get out," she told a reporter.[59] The problem was simple economics: She just couldn't afford to leave.

St. Charles farmers had the opposite problem: They couldn't afford to *stay*. By the 1990s, the levees in the triangle had an effective protection of 11 to 13 years.[60] In other words, farmers could expect their land to flood once every decade or so, causing them economic disaster.

Many farmers also believed that levees built to protect commercial development in neighboring St. Louis County only aggravated their flood problem, though the corps, which approved those projects, denied this with all its bureaucratic heart.[61] In any event, County Executive Eugene Schwendemann pleaded the farmers' case after the 1993 flood, arguing for a 50- to 100-year structure. He also wanted assurance from nearby communities both up- and downstream that they would not build their levees higher lest St. Charles continue as a "dumping ground."[62]

Meanwhile, Ehlmann, now a state senator, and a local state representative named Joseph Ortwerth brought pressure on the U.S. Congress to get FEMA to lighten up on its regulations. Their chief concern was the 50-percent rule. Under this regulation, if an insured structure in the floodplain is more than half damaged, it must be torn down and rebuilt one foot above the 100-year-flood level. Ehlmann and Ortwerth wrote the entire congressional delegation from Missouri, crying foul. "While this requirement may make sense with regard to clubhouses, trailer parks and other structures built inappropriately on flood-prone properties, it is patently unfair to apply this rule to established communities and farmsteads which never before experienced flooding prior to this summer's extraordinary incident." Welcome to the self-serving world of floodplain politics. According to Ehlmann, FEMA simply did not understand that some people—the farmers of course—*had* to live in the floodplain. Then there were those, the poor, who didn't have to be there but did so anyway, of their own free will: "They live there because it's scenic or the rent is cheap."[63] After all, it's not every day that people get to *choose* to live in an utter cesspool.

Ehlmann and Ortwerth got their way. Although FEMA kept the 50-percent rule, it made a significant change in how structural damage would be calculated, effectively subverting the regulation. The government allowed homeowners to use replacement cost as opposed to market value in figuring whether the building had been damaged by more than half. The result was obvious. It became far less likely that a structure would be condemned if the damages were calculated as a percentage of the much higher replacement-cost figure. The 50-percent rule was now more like a 75-percent rule or higher. People barreled into the St. Charles County Planning Department with requests for permits to rebuild. According to Darren May, a planner at the agency, at least 80 percent of those who came through the doors of his office arrived with estimates to repair damage of 48 to 49.9 percent. He saw estimates for new carpet at $1 per square foot—including the pad and labor! He saw carpenters billing out at $4 per hour.[64] St. Charles County was once again doing what it did best: setting itself up for more disaster, with the help of Ehlmann, Ortwerth, and the staff at FEMA.

DEATH OF THE RIVIERA

Today, Missouri Mobile and the two other mobile home parks near the Missouri River are gone. You won't hear Steve Ehlmann complaining. In the 1992 election, the precinct with the parks delivered him just one-third of its votes, his worst showing. With the parks bulldozed, Ehlmann can rest easy about his electoral prospects. Asked where the people who once lived there have gone, Ehlmann replied, "I don't know."[65]

What of the mobile homeless and other poor homeowners displaced by the flood? Many simply were forced to leave St. Charles County, where affordable housing was in very short supply even before the water rose. According to Shelia Harris-Wheeler, who chaired a group aiding some 2,800 homeless families, nine out of ten people were relocated to other counties, such as Lincoln, Warren, and north St. Louis.[66] Many now wound up commuting much greater distances to jobs—if they had them—in the St. Louis metropolitan area. They also lost whatever community they once had in their old neighborhoods. Simply put, the disaster drove the poor out of St. Charles. Miriam Mahan, the director of St. Joachim and Ann Care Services, explains that what happened after the flood succeeded in "what we call blocking the pursuit of happiness."[67]

Of course, it wasn't exactly a bad idea for people to move out of the floodplain, and the federal funds provided to pursue this goal were seen by many as one of the great benefits to come out of the 1993 disaster.[68] Only many mobile home owners in St. Charles were sent packing after being given the rawest deal possible. This was especially true of those who rented lots in the mobile home parks. Under Missouri law, mobile homes were personal property, not real estate. And as far as the state was concerned, there was no difference between living in your mobile home and living in your car. Since federal funds could not be used to purchase personal property, those who rented sites received no buyout money. Nor did they receive relocation funds. The owners of the mobile home parks were of course paid handsomely for their land; the renters, people such as Scott Gluckhertz, got nothing. Gluckhertz wrote President Clinton a letter noting that the owners of conventional housing received buyout money but not he and his fellow mobile home owners, who rented land for their trailers. "What is the difference between other homeowners and mobile-home owners when it comes to the cost and inconvenience of relocating?" Gluckhertz asked. He had a point. The government's policy amounted, in his mind, to "an unfair form of discrimination toward lower-income individuals."[69] Jerry King, a consultant hired by the county to oversee the buyout program, explained that even though the program was "a step forward" in reducing the population at risk of flood, it was "a giant step backwards" in that it further reduced the availability of affordable housing in a county already suffering from a significant shortage.[70] The best thing you can say about the mobile home buyout is that when the 1995 flood struck—its crest of 36.5 feet making it among the worst floods in county history—some 2,500 low-income people were out of harm's way. Not that their lives were necessarily any better. But they weren't wet.

At one point, a touch of humanity materialized. County officials came up with a plan to use federal relief money to build a new subdivision of affordable housing in a safer spot not too far from where the old mobile home parks existed. That way residents could return to some semblance of their former lives. It was the decent thing to do. But when residents of nearby subdivisions got wind of the proposal they were furious. Not in my backyard, they yelled. What, subsidized housing here? More than 400 such residents signed a petition opposing the project.[71] The plan fizzled. In 1994, another plan emerged to build affordable housing further north in the county, but nothing came of that plan either.[72] Even though FEMA's buyout program had precipitated the need for alternative housing, the agency did not aid local planners in dealing with the problem.[73]

It is always nice to have a happy ending. Which brings us to the farmers and their flood woes. At last report, it looks as if they are going to get their levee. In 1996, Congressman James Talent (R-Missouri) shepherded legislation—written largely by Ehlmann—through Congress that allows the levee district to at least take the initial steps in building to a 20-year-flood height without jeopardizing the county's status in the flood insurance program.[74] It is only a two-decade levee, of course, and will never be as tall as the farmers would like. But it is a levee all the same, and a foundation on which to pile more dirt some day. If the farmers load the dirt on high enough, maybe they can shield themselves from any reminder of the pain and suffering that displaced people lower down the social ladder experience as they make the long commute from their mobile homes in, say, Lincoln County to work in St. Louis.

INTERLUDE

....

The Perils of
Private Property

Let me take a moment to sketch in some of the broader trends in the history of natural disaster during the postwar decades. As our journeys to Florida and Missouri suggest, the shift to fewer deaths paralleled a rise in property damage in the period since 1930.[1] Of course, there was a lot more property *to* get damaged, no matter where it was built. Nonetheless, in recent years natural calamities have proved more financially devastating than ever. Figures for insured losses in all U.S. catastrophes—from riots to hurricanes and earthquakes—reveal that the 10 costliest were all natural disasters.[2] Moreover, all have occurred within the last decade, a period, ironically, that Congress along with the United Nations has officially dubbed the Decade for Natural Disaster Reduction![3]

It's no wonder that the nation's insurance executives are worried. "The fact is that the 20 years prior to Hurricane Hugo [1989] was kind of a vacation," explains the vice president of one insurance company. Property insurance "used to be a nice, safe little line."[4] But this state of affairs has changed dramatically in the 1990s. Just two disasters, hurricanes Andrew and Iniki, drove nine insurance firms out of business.[5] Not since Hurricane Camille in 1969 (a category 5 storm, the most powerful with winds in excess of 155 mph) had the United States experienced a severe hurricane.[6] That 20-year lull—an anomaly, historically speaking—only encouraged more population growth and development in coastal locales, aided by insurance companies eager to sell more policies.[7] Now insurers are finding themselves overexposed. Before 1992, many insurance experts foresaw no more than $8 billion in damage from even the worst windstorm. In fact, the insured losses from Andrew totaled over $15 billion—largely the result of the unexpected and spiraling costs of rebuilding. Most worrisome of all, the Florida hurricane, which did its worst damage south of the densest concentration of people and property in Dade County, was, according to one study, just "a wobble away" from being a $65-billion storm.[8] In this respect, the disaster was hardly the Big One that the media made it out to be.[9]

Is the recent shift toward higher levels of damage the product of more violent natural conditions? It's unlikely. A recent study of intense Atlantic hurricanes, for example, has shown that, in fact, over the last half century, the incidence of such powerful storms has actually *declined*.[10] Yet as Roger Pielke and Christopher Landsea have observed, many—especially those in Congress, insurance, and the media—still attribute the recent high damage levels to an increase in hurricane activity.[11] Moreover, the tendency to blame this fictional increase in hurricane incidence on changing climatic conditions has further naturalized the problem. Hurricanes, proclaims a U.S. Senate task force on disaster relief, "have become increasingly frequent and severe over the last four decades as climatic conditions have changed in the tropics."[12] But as Landsea and others point out, the high damage costs are not the product of climate change. They stem instead from increasing vulnerability to hurricane disaster as population and development have forged their way in along the coast.[13] In other words, "man-made" changes, not natural ones, are responsible for the increased destruction.

Casting nature as the villain—as many still do—presents more than just an interpretive problem. This mind-set has had practical effects. Invoked to justify the use of more technology in the fight against natural calamities, the erroneous assumption that nature is the main culprit has doomed many such technical solutions to failure. Consider for a moment flood losses, which have grown over the twentieth century despite a steady and generous expenditure of government resources aimed at stemming them. Flood control, once largely a local concern, first became a major federal responsibility in response to the devastating inundations of the 1920s and 1930s. The government's program, as implemented by the U.S. Army Corps of Engineers, initially amounted to a massive containment strategy. Seeking to subdue unwieldy nature with all the concrete and steel it could muster, the corps by the 1960s had spent some $9 billion building over 200 reservoirs, not to mention almost 7,500 miles of channel modifications and more than 9,000 miles of levees and floodwalls.[14] Despite this technological onslaught, flood losses continued to mount. Even with an investment of some $25 billion over the decades, per capita flood losses, adjusted for inflation, were almost two and a half times greater in the period from 1951 to 1985, than between 1916 and 1950.[15]

With the bankruptcy of the containment strategy abundantly clear by the late 1960s, government officials shifted toward the nonstructural approach embodied in the National Flood Insurance program. Yet, as already noted, flood insurance itself is not without major problems, the worst involving the failure to adhere to the spirit of the 1968 law. Thus, in response to the devastating midwestern floods of 1993 ($12 billion to $16 billion in damage), policymakers again headed back to the drawing board. At the heart of the latest approach is watershed management. This is a broader nonstructural tactic meant to encourage wetlands restoration and relocate people out of harm's way.[16] In the spring of 1994, for example, the town of Chelsea, Iowa, voted to move to higher ground in connection with a federal government plan to get people out of floodplains. "It used to be that we managed the river and let people go where they want to," remarked the town's mayor, Rodney Horrigan. "Now we are trying to manage the people and letting the river go where it wants to."[17]

At face value, this seems like a sensible alternative to the subsidization of risk. But relocation is an intensely political issue that raises many questions of practice and equity. While moving may be a realistic option for small communities, larger cities that owe their existence to elaborate water control structures are unlikely to find it appealing. "We can't pick Des Moines up and put it on a hill," explains Harry Kitch of the Army Corps of Engineers.[18] For a variety of reasons, there will probably always be people who refuse to move, out of a concern for money, a love of place, or sheer stubbornness. Dismantling the complex set of state-sponsored risk taking that has grown up over the decades is like trying to disassemble a house of cards, and who will pay to take those cards apart is a sticky question. Earl Bull, a farmer from near an Illinois town flooded in 1993, asks: "If we're going to force farmers out and move people out of harm's way in the Midwest, are we going to do that as national policy? How about the San Andreas fault?"[19] It's not a bad question for policymakers, bent on undoing years of institutionalized risk taking. The San Andreas aside, no one, after all, is calling for a buyout of the Hamptons or any of the other eastern beachfront haunts of the rich and famous, which are potentially at risk from another earthquake like the one that struck Charleston in 1886.

In fact, the reverse process is occurring along the nation's coasts, as law and politics conspire to encourage the stubborn entrenchment of the well-off. Isle of Palms, a six-mile-long barrier island in South Carolina, is a case in point. Isle of Palms remained largely undeveloped until the 1940s, when J.C. Long, a Charleston lawyer, acquired a major portion of it. Long, who went on to become one of the state's most prominent developers, set his bulldozers to work leveling dunes—some as high as 30 feet—in order to make way for residential construction.[20] Destroying the dunes, which absorb some of the energy of storm surges and act as a front line of defense against the sea, only made the island more vulnerable to hurricanes.[21] But with the exception of Hurricane Gracie (category 3) in 1959, the South Carolina coast remained largely free of major storms until 1989, when Hugo struck. The lull in hurricane activity only encouraged further development of the shore. In the 1970s, developers transformed the eastern end of Isle of Palms into a resort, later named the Wild Dunes Beach and Racquet Club, replete with expensive housing, tennis courts, and one of the world's finest golf courses, all neatly locked away behind a security gate.

Enter David Lucas. In 1986, Lucas, a builder and developer, bought two oceanfront lots in Wild Dunes for close to $1 million. On one lot he planned a home for himself; on the other he envisioned what he called "a speculative house."[22] Only he never was able to build. Two years after he made his purchases, South Carolina passed a beach management act. By the 1980s, some state legislators worried that increasing development, especially the rising number of seawalls, threatened to destroy the state's valuable beaches, thus compromising the region's tourist potential. So a set of regulations was drawn up—by no means a stringent set—to help limit coastal development.[23] The new law prevented Lucas from realizing his so-called dream home (or the speculative one, for that matter) because it deemed that building so near the shore would pose a hazard. Lucas subsequently sued, arguing that an unconstitutional "taking" of his property had occurred. A lower court found for Lucas, but the South Carolina Supreme Court overturned this decision, clearing the way for the U.S Supreme Court's ruling in 1992, which we will turn to in a moment.[24]

A visitor to the Lucas lots in the early 1990s would have found the land dry and suitable for building. Why not build here on the shore, with all it has to offer in serenity and beauty? For starters, the land was covered with water four feet deep when Hurricane Hugo (category 4) struck in 1989. In 1983, erosion so threatened the homes nearby Lucas's lots that they required emergency sandbagging. In the 40 years before Lucas made his purchase, the Atlantic Ocean had completely flooded his land no less than four times.[25] In other words, Lucas's dream might actually have turned into a nightmare if strong wind and water had ripped his home apart and sent pieces crashing into neighboring houses.[26] But natural disaster probably was not the first thing on Mr. Lucas's mind, because Isle of Palms is part of the National Flood Insurance program. Let disaster strike and the federal government would soon be cutting him a check to rebuild.

In its 1992 opinion, the U.S. Supreme Court ruled for the first time that a land-use regulation draining all the economic value from a piece of property constitutes a taking and necessitates compensation. South Carolina had to pay Lucas for his land unless it could prove that by building he would be creating a public nuisance. In other words, the state's decision against compensation could not simply rest on the general notion of serving and safeguarding the public interest. Some observers believe the case will have limited impact because it involved an unusual situation in which a taking deprived the land of *all* of its value. But justices Harry Blackmun and John Paul Stevens, who dissented, feared broader implications. According to Stevens, the ruling might "greatly hamper the efforts of local officials and planners who must deal with increasingly complex problems in land-use and environmental regulation."[27] Although the case pertains only to a total loss of value, the court's reasoning could be applied to a future case in which the land is less than 100 percent compromised—which is no doubt an exciting prospect for developers like Lucas and others interested in getting the government off their backs.

The Supreme Court's ruling in *Lucas* was followed two years later by another case that, if anything, provides yet more ammunition for those hell-bent on doing what they wish on their land, even if it means flirting with disaster. In the *Dolan* case, decided in 1994, an Oregon community required a store located on a floodplain to set aside 10 percent of its land for drainage, green space, and a bicycle path, in return for the right to expand and improve the property. The Supreme Court ruled that the land code requirement constituted a "taking" under the Fifth Amendment unless the city could show a "rough proportionality" between what it was requiring the store to do and the projected harms of the store's proposed expansion. In a critical shift, the court placed the burden on the city to justify the land-use restriction; in the past, that burden had fallen on the landowner, who had to show that a regulation removed all the value from the property. "Property owners have surely found a new friend today," remarked Justice Stevens, who dissented.[28] And if the law allows property owners like the Dolans more latitude to develop their land, it seems almost certain that the result, in hazardous environments at least, will be self-made disaster.

If *Lucas* and *Dolan* are any indication, the trend is toward making private property even more private, a welcome development from the perspective of those, like Lucas, involved in the surging property-rights movement. Some 600 property-rights groups have sprung up in the United States since the late 1980s. Dedicated proponents of private property, these organizations stand opposed to virtually all environmental and land-use regulations—one of the main lines of defense against natural disaster, however imperfect. In 1997, they found support from Republicans in Congress who tried to make it easier for developers, fed up with zoning and other government land-use restrictions, to sue in federal court.[29] Private property owners across the nation, including those occupying some of the most dynamic and hazardous places around, seem almost destined to be handed another shovel as they dig in still deeper on their land—a move at direct cross-purposes with other government attempts to retreat from underwriting life in disaster-prone locales.

This is not to say that all building is bad. Nor to advocate a retreat from San Francisco, Missouri, Florida, or even Isle of Palms. The point, instead, is this: When we build we need to do so with our eyes open, to be clear about who wins and who loses in governmental efforts to control and predict natural disaster (subjects addressed in the next chapter), clear about who, if anyone, is going to pay for calamity when it does strike, clear about the complex interaction between human economic development and natural forces, and clear about the extraordinary public costs involved in protecting private development from itself.

III
CONTAINING
CALAMITY

SIX

....

The Neurotic Life
of Weather Control

In 1954, Howard Orville, a retired navy captain and advisor to President Dwight Eisenhower, broached a plan for "complete weather control," a scheme designed, effectively, to put the atmosphere on a steady dose of tranquilizers.[1]

Those were heady days, the 1950s. Predicated on the illusion that nature remained largely to blame for the natural disaster problem, weather modification ranks among the grandest schemes on record for warding off calamity. It came of age during the cold war at a time when many American political leaders were obsessed with eradicating evil forces, be they communist spies or disorderly weather patterns. No single technological intervention better demonstrates the abortive—at times bizarre—logic at the core of the dominant approach to natural hazards.

Weather modification as it exists today was the brainchild of a self-trained chemist and high school dropout named Vincent Schaefer. In 1946, Schaefer, a clean-cut sort who worked for General Electric at the time, climbed aboard a plane and dropped six pounds of dry ice into a cloud near Schenectady, New York, making it snow.

The following year, the military teamed up with Schaefer and his colleague, Irving Langmuir, a Nobel laureate in chemistry, to tackle a hurricane threatening the Florida coast. Planes dumped almost 200 pounds of dry ice into the storm's eye. A few hours later, the hurricane shifted course and plowed smack into Savannah, Georgia, causing some $5 million in damage. Langmuir believed the seeding made the storm change track, holding out hope that future hurricanes could be steered away from populous areas. Although most atmospheric scientists disagreed with that view, that day, ironically, Savannah, an insular city steeped in history and tradition and incredibly resistant to change, may have been on the receiving end of modernity ascendant: the first engineered hurricane disaster.[2]

In any case, weather modification, feeding on fantasies of a free market in climate, boomed in the wake of Schaefer's initial success. And the later discovery that silver iodide could be used effectively to seed storm clouds only furthered such pluvial ambitions. By the early 1950s, roughly the time of Orville's modest proposal, commercial cloud seeders were off tinkering above almost 15 percent of the country, mostly to alleviate droughts in the West.[3] Even an industry trade group emerged, the Weather Control Research Association (later renamed the Weather Modification Association), claiming that apart from hurricanes and droughts, cloud seeding had the potential to neutralize the effects of hail, fog, and lightning. Among the group's corporate members were the Los Angeles Department of Water and Power, various electric utilities, as well as private weather companies with names like Atmospherics and Better Weather Incorporated, all of them fierce supporters of the new capitalist weather ethos.

Meanwhile, a spate of urban tornadoes in 1953 that roared through the North hitting Flint, Michigan (116 dead), Worcester, Massachusetts (90), and Cleveland, Ohio (19)—not to mention a vortex siting in, of all places, the Carnarsie and Flatlands sections of Brooklyn, the very essence of urbanization—galvanized weather technocrats in a way that the deaths of poor rural blacks and so-called trailer trash had never succeeded in doing. At the U.S. Air Force Missile Test Center in Patrick, Florida, plans were put forth before year's end to use atomic warheads to destroy tornadoes lest they creep up any further on the nation's metropolises.

Freighted with all the hallucinatory appeal of the latter-day Star Wars scheme, the twister project was also shot through with irony. The 1953 tornadoes, as it turns out, were suspected by some as resulting from recent atomic explosions, a fear that even prompted one physics professor from Holy Cross College in Worcester to take a Geiger counter to a baseball-sized hailstone delivered up in the storm (result: negative). In spite of such concerns, the military—with the iron law of conservation of atomic angst as its guide—rushed headlong into nuclear weather war, assured by none other than Vincent Schaefer, the father of cloud control, that the A-bomb, in all likelihood, had little to do with the tragedies.[4]

The macho thinking behind Twister Wars and other such futuristic dreamwork was nicely captured by Adm. Luis de Florez, like Orville, a retired navy officer, in a 1961 speech titled "Weather—Take It or Make It." "We have not tackled the control of weather with the same determination and audacity that we have shown in harnessing nature in other ways," he remarked, annoyed by the federal government's ultimate failure to commit to a large-scale weather modification program. "It is strange, indeed," he observed, "that the American people...display the same fatalism and resignation about the weather that our remote ancestors did thousands of years ago. It is a fact, nevertheless, for when it comes to weather, we accept whatever comes passively, patiently, as an act of the Gods."[5] But de Florez and his fellow weather makers were unwilling to take no for an answer in their dealings with nature. Theirs was a mental universe where the domination of the natural world had been elevated into a moral absolute, where acts of God were pathetic relics of a premodern past. Ray Jay Davis, a lawyer and president of the Weather Modification Association in the 1970s, put it this way: "I believe nature is not the image of God; it is his handiwork. In this world, as Eric Hofer [*sic*] puts it, 'anyone who sides with nature against man...ought to have his head examined.'"[6]

Also weighing in for *Homo sapiens* was Edward Morris. A lawyer and president of the Weather Modification Association in the 1960s, Morris had successfully defended a California electric company against charges of cloud seeding a flood into existence. "Whether or not we all like it," he wrote, "large scale weather modification lies in man's future." Seeking to further naturalize attempts to control the weather, Morris pointed out that "unintentional" weather modification through industrial activity and the release of carbon dioxide and other pollutants was currently taking place. With weather modification already occurring, albeit unintentionally, what harm could exist in purposefully trying to alter the weather? Only one problem remained: what to do with those opposed to such an approach. "To overcome the emotional opponents of weather modification," Morris explained, "it will require more than a letter from a government agency that its files contain no evidence of disasters caused by cloud seeding." It was going to take a federal agency organized along the lines of the Atomic Energy Commission or the FCC, Morris explained. "This Federal agency might eventually plan weather by zones or by days, sell or buy weather to or from other agencies, possibly even make decisions as to the 'best' weather for certain times and places."[7]

As it happened, the idea that cloud seeding could make the weather into a commodity to be bought and sold like anything else in America was more a daydream than a real possibility. Although Schaefer and Langmuir showed that clouds could be modified, the efficacy of this technique at increasing precipitation itself remains very much in doubt. A half century after the technology first emerged, only two studies—conducted in Israel in the 1960s and 1970s—claim to show a statistically valid increase in precipitation. And even these studies have been roundly criticized for failing to demonstrate significant results.[8] In short, cloud seeding is the ultimate in pseudoscience. (The lack of statistical support was a point of no small annoyance to Vincent Schaefer, who called the misuse of statistics to assess cloud-seeding's efficacy a "major mistake."[9])

Nonetheless, from the 1960s through the mid-1970s, cloud seeding remained high on the agenda of a number of policymakers, weather entrepreneurs, and congressmen (especially those from the arid West and the hurricane-prone areas of the East). In 1961, government support increased as the U.S. Bureau of Reclamation, led by the incorrigible river dammer Floyd Dominy, began Project Skywater. The bureau—whose primary responsibility, historian Donald Worster notes, centered on the domination of nature, had, since early in the century, catered to the water needs of wealthy farmers in the West.[10] It thus made perfect sense that an organization so dedicated to conquest would step in to sponsor (after the U.S. Weather Bureau refused to do so on scientific grounds) a technocratic dream for further augmentation and rationalization of the nation's water resources.

But a strong commitment to conquering nature was hardly the domain of the Bureau of Reclamation alone. In 1962, the departments of

Defense and Commerce initiated Project Stormfury—a cloud-seeding program bent on sending planes off into the Atlantic to neuter hurricanes. If nothing else, the publicity surrounding hurricane seeding in the 1960s laid bare the moral uncertainties of this new technology. Although Project Stormfury aimed to reduce wind speeds, some believed that cloud seeding still might be used to cause storms to shift course. The ability to steer hurricanes would, of course, have had a major effect on the insurance industry—a point that prompted two analysts from the Travelers Company to concoct the following science fiction scenario. What would have happened in 1964, they wondered, if Hurricane Cleo had been nudged 10 degrees off track? The answer: "Hurricane Cleo, which caused more than $65 million insured property damage, could have missed Florida entirely and its land entry postponed until it reached the vicinity of Georgetown, South Carolina—a relatively unpopulated area." All this sounds great unless you happen to live in Georgetown, a poor area with a large black population. Of course, the analysts could have moved Cleo slightly to the left of its natural track. But this was not done, they explained, since the hurricane would still have wound up crossing paths with Florida, which had long since overdosed on development.[11]

For the weather modifiers, the moral issues were largely irrelevant in their crusade to control nature—a mission that, to them, seemed as natural as it was inevitable. In the 1970s, the booster Weather Modification Association published an informational pamphlet on the "facts" about cloud seeding. Readers may have been surprised to learn that there is no such thing as natural weather, at least not anymore. "In the sense that 'natural weather' means cloud formations, storm systems, and precipitation which are unaffected by human beings, there has been no 'natural weather' since the first fire was intentionally ignited by early man." Translation: Cloud seeding was as natural as striking a match. Nor was there any need to worry about God or morality. "Is cloud seeding against God's will?" the pamphlet asks. Surely not. "Cloud seeding is a tool to be used by mankind just like a tractor, airplane or space vehicle."[12] But to portray cloud seeding as a neutral tool obscured the social relations embedded in this technology. Cloud seeding was not simply an instrument, but a set of values and moral assumptions about what kind of world should exist and who would benefit from it.

One question the pamphlet did not address: If no such thing as natural weather existed, was there no such thing as "natural" disaster either?

Who, if anyone, would be held responsible for disaster in this post-modern world, where the line between the natural and the engineered had seemingly disappeared?

WEATHERGATE

Which brings us to drought, a topic not discussed as yet. But we need briefly to explore it if we are going to understand fully the moral conundrums at the heart of cloud seeding. The drought at issue scorched the Northeast in the early 1960s and emerged as one of the most devastating dry spells in American history.[13] Many areas were drier than at any time since the nineteenth century, causing the federal government to declare farm disaster areas in a record number of counties. It was so dry that apples ripening in New England barely achieved the size of golf balls. In New York City, fountains were turned off and glasses of water were no longer served, unrequested, in restaurants. Tiffany's substituted gin for water in its window display. A Manhattan restroom hung out a sign: DON'T FLUSH FOR EVERYTHING.

The drought lasted four long summers, from 1962 through 1965. One particularly hard-hit region was Fulton County, Pennsylvania, a desperately poor, rural area tucked away in the south central part of the state. There, amidst all the heat, dust, and dried-up dreams, we find a group of dairy farmers who suspected that God alone was not behind their severe weather woes.

Their reasoning seems clear enough. In the late 1950s, a group of well-off commercial fruit growers, desperate to stop the damage that hail caused their fruit crop, launched a cloud-seeding program. (Because they relied on irrigation systems, drought proved less of a concern for the fruit farmers.) After the failure of this initiative, the fruit farmers (dispersed in the states of Pennsylvania, Maryland, Virginia, and West Virginia) banded together and hired W.E. Howell Associates to chase the hail from their skies. For two consecutive summers, beginning in 1963, Wallace Howell, a Harvard and MIT graduate and a leading figure in the weather modification business, vaporized silver iodide over the Blue Ridge region in an attempt to stop the hail that mercilessly pounded the farmland, hail that at times could reach the size of baseballs.[14] Nobody can say for certain what effect Howell and his company had on the weather. But according to the dirt farmers excluded from the scheme, the cloud seeding succeeded all too well, putting an end to all precipitation whatsoever.

Farmers in Fulton and Franklin counties suspected the cloud seeding as the source of their blistering weather as early as 1962. That year vandals cut down over 100 plum trees on land owned by the Heisey Orchards, one of the main sponsors of the seeding. In an effort to quell the unrest, the Heisey company had its lawyer invite Charles Hosler, the head of the department of meteorology at Penn State University, to a meeting with some 400 irate farmers. Hosler himself had opposed commercial cloud seeding since 1950. He told the farmers that despite the expenditure of some $70 million over the course of 17 years, weather modification amounted to one big failure. "Not in one place has it been shown that cloud seeding has any effect on the weather."[15] Remarks such as this, designed to quell the fears of farmers, made many wonder if Hosler was simply a hired gun for the large agricultural interests. Indeed, the Heiseys were paying his way, though Hosler later claimed that he did not discover this until after he arrived to speak. Nor presumably did he realize until he got there (and found plainclothes police on hand to protect him) that he was risking his life by attending.[16]

Wallace Howell, for his part, saw the dairy farmers' opposition to cloud seeding as simply the product of one big high-pressure system. In theory, the farmers' objections to weather modification rose and fell in inverse proportion to the rainfall: the more rain the less opposition, and vice versa.[17] But, in fact, no such simple equation can be used to describe the opposition, which ran along much deeper and more complicated lines. Simply put, the farmers fought weather modification because it offended their moral sensibility and their customary sense of how nature should be used.

The views of Guy Oakman, the head of the Pennsylvania Natural Weather Association (PNWA), which was formed in 1964 to fend off the cloud tampering, are a case in point. Oakman believed all landowners had a right to the natural weather over their land. "We fail to see how anyone, with a just regard for the rights of a property owner, can ever hope to apply the science of weather modification without infringing on these inherent rights."[18] The idea that someone, an outsider like Howell no less, could come into his community and meddle with his weather was to Oakman and others opposed to seeding clouds a moral outrage. Forget the drought, Oakman seemed to be saying. "We believe the question is much broader and more basic," he explained. "We believe the question is: Does a man have the right to have the weather above and around him remain in its natural condition and undisturbed? We believe he does."[19]

Yet more than a concern with property drove forth the natural-weather people in their crusade. Cloud seeding, they also believed, trampled on God's will. "Since man can sometimes be very selfish," wrote Oakman, "we believe, that the control of the clouds, the rising and setting of the sun or the four seasons should be left to a higher power."[20] If God sent a drought, storm, or some other weather extreme, then presumably there were good reasons for its appearance and, in any case, everyone would suffer equally, surmised Oakman and his comrades. Said Marl Garlock, a Pennsylvania farmer and supporter of natural weather: "From the beginning of time until recently, the weather has been controlled by Our Creator. Why then should we have men professing to control our weather today, to suit the whims of a few people, with so many left to suffer so much."[21] For the dairy farmers, acts of God were supposed to be just that: events for which no mere mortal could be held accountable. To tamper with the weather and possibly create disaster challenged the word and the will of the Lord. As Delmar Mellott, a farmer and a member of the PNWA said, "Those who talk of the practice of weather control in any way should take note to Psalms 46:10: 'Be still and know that I am God.'"[22] At another point, Mellott explained further: "I feel that if there has to be talking to a—what I call the rain maker, if I call for rain, I would rather do it on my knees, not on the telephone."[23]

At the core of these objections lay a deep-seated concern for the preservation of fundamentalist calamity. Driven by their own particular version of moral rectitude, the self-professed victims of this technological hubris felt compelled to question, as Oakman put it, "whether any person or group of persons has the right to interfere with the natural weather phenomena."[24] For these latter-day weather Luddites, natural calamities were acceptable as long as they remained natural and moral—nonrandom events caused by a just God. But to the dairy farmers' minds at least, the drought was a different story altogether. In contrast to the dominant view of the disaster as a chiefly "natural" calamity, the natural-weather people operated within a completely different causal framework. This disaster did not simply happen, they said. It occurred, instead, when some people decided to play God with the atmosphere, blurring what they took to be the hard, crucial line between the engineered and the natural.

That line became even fuzzier as cloud seeding continued in the 1960s, raising the specter of other seemingly unnatural disasters. In 1969 Hurricane Camille, the second most intense storm in U.S. history (behind Florida's Great Labor Day Hurricane), ripped across the East, killing 256 people and causing over $1-billion worth of property damage. Who or what could possibly be responsible for such a ferocious storm? The choices: (a) nature, (b) big-time agriculture and its hired weather cowboys, or (c) the government of the United States of America. Most, no doubt, opted for nature, but not the Tri-State Natural Weather Association (formed to combine all the various opposition groups that had sprung up after the drought), which placed the blame squarely on the shoulders of Uncle Sam. In a brief but hard-hitting publication titled *Cloud Seeding: The Science of Fraud and Deceit: (A Criminally Conspired Complex)*, the group claimed that Project Stormfury had backfired, causing Camille to spin wildly out of control. "The people who died from this hurricane were killed by scientific blunderings making those who tampered with this storm absolute murderers."[25]

Despite such claims of evildoing, there is no evidence that Camille was seeded. The Project Stormfury staff did, however, conduct weather modification operations the very next day (August 18, 1969) on Hurricane Debbie. A substantial reduction in wind speed inside the hurricane made this one of the most promising storm modifications ever. At the time, Project Stormfury protocols prohibited seeding any tropical depression likely to make landfall. But some scientists, entranced by the apparent triumph over Debbie, later urged the government to lift the ban in order to allow operational seeding of hurricanes threatening populated coastal areas.[26]

Although the prohibition remained in place, the natural-weather people remained unconvinced, implicating seeding in the destruction wrought three years later by Hurricane Agnes. Agnes killed 117 and caused over $3 billion in damage, but again, as far as anyone knows, no storm modification took place.[27] The storm caused some of the worst flooding in Pennsylvania history, prompting Paul Hoke, a farmer from Franklin County and the president of the Tri-State Natural Weather Association, to decry the emergence of a pattern of "crazy weather." Hoke and his organization again singled out Project Stormfury for derailing the hurricane from its natural path and causing it to wheel out of control. And since Congress appropriated the money for such seeding

This is the internal capture of the U.S.

The proposed Potomac River Basin Compact —
a new government of America,
operated by citizens from another country — by 1975!

The Master Land Plan of America – already in place
whose hidden effect is,
CONCENTRATION CAMPS behind BARRIERS.

**Tri-State Natural Weather Association literature
(Pennsylvania Department of Agriculture)**

projects, Agnes constituted, as they wryly pointed out, not an act of God but "an act of Congress."[28]

When it was not trying to derail storms, Hoke and the Tri-State Natural Weather Association contended, the U.S. government busied itself seeding clouds in the Potomac Valley near their homes in an attempt to engineer a drought. Why in the world would the government opt for something so sinister? As the natural-weather people argued, the government wanted to drive residents out of the valley so the U.S. Army Corps of Engineers could then purchase the land on the cheap. Then it could build dams and control the waters of the valley as it pleased. This was all presumably part of a plan that would lead, so they claimed, to the "internal capture" of the United States. In place of the 50 state governments, the anti–cloud seeders believed that a master plan existed to break up the country into 22 river basin systems. The nation would be reconfigured politically and geographically and run by river basin commissioners "whose members will have Americanized names, but who will be citizens from another country."[29]

In the minds of the natural-weather people, cloud seeding amounted to one huge conspiracy, and it was wrong. It was wrong because it went against God's will. It was wrong because it offended against morality. It was wrong because it was antidemocratic and threatened to wrest control away from the people "with the eventual hope of enslaving them."[30] It was wrong because it was a form of pollution involving toxic substances such as silver iodide. And it was wrong for ecological reasons: "Natural weather is the only way mankind can live happily and in biological compatibility with his environment."[31] Despite such obvious wrongness, the natural-weather people claimed, seeding continued to be practiced, and with terrifying results.

The continued seeding finally drove one of its most vocal opponents out of business, or so he said. In 1978, Paul Hoke announced his intention to leave farming to work in a plumbing store: "I'm quitting farming because I can't raise crops, and I can't raise crops because of cloud seeding."[32]

If the opponents of cloud seeding sound a bit weird and paranoid, there was certainly no shortage of critics willing to point this out. Many dismissed them as cranks and crackpots. Richard Heisey, a commercial farmer instrumental in bringing seeding into the Blue Ridge area, said this about those who attacked weather modification: "Ninety percent of the people who were up in arms were, caliber-wise, mentally, ah...Well, let me tell you, there are people in this area who think that when man landed on the moon it was fake. They think it was filmed in the desert." Added Pennsylvania Agriculture Secretary Kent Shelhamer, "You have to remember that farmers tend to be suspicious and sometimes have a persecution complex."[33]

Had the renowned historian Richard Hofstadter discovered the natural-weather people, he no doubt would have pressed them into service in his essay "The Paranoid Style in American Politics." Paranoid thinking, explained Hofstadter, often germinated in the ashes of catastrophe.[34] Add to this a concern with persecution and a weakness for conspiracy theories and the result was a persistent tendency to look over one's shoulder. Hoke and his associates seemed to qualify on all counts. In the 1970s, the natural-weather people even went so far as to call cloud seeding the equivalent of "meteorological Watergate."[35]

Is it paranoid to see droughts, hurricanes, floods, and other so-called natural disasters as the product of unnatural, human acts? Are such claims merely the ravings of a bunch of poor farmers pushed over the edge by foreclosure?

Alongside the paranoid style in American politics, we must make room for another equally important cultural tendency: the diagnostic style. This is the tendency to offer a clinical diagnosis in place of empathic understanding. Instead of trying to understand why people might oppose a dominant set of political or, in this case, scientific and technological ideas, the diagnostic style proceeds immediately to the realm of nosology to discredit such individuals, transforming their opposition to the cultural mainstream into a form of mental illness. Occasionally, the diagnostic style winds up condemning people for thoughts that, if delusional, are probably no more out of touch with reality than the idea—a favorite of some seeding boosters—that with enough technological muscle we can rid the world of natural disaster. The reasoning of the natural-weather people is probably no worse than those who persisted in a collective scientific wish for total control over natural extremes despite virtually no evidence supporting the efficacy of cloud seeding. In short, if the natural-weather people were deluding themselves, then so were their opponents. Ultimately, cloud seeding reveals a great deal about the assumptions of the prevailing scientific view of calamity. That understanding sees nature—as opposed to, say, human economic forces—as the real problem. Having constructed natural disaster as chiefly the result of geophysical extremes, the scientific community is then consulted by politicians eager to rationalize the expenditure of large amounts of money in an effort to combat nature's fury.

Oakman, Hoke, and their followers, however, were articulating a view of calamity that, while admittedly radical, has a certain logic. Of course, their understanding seems ridiculous to a culture bent on portraying catastrophes such as droughts and storms as morally neutral, natural phenomena, well beyond the pale of human action.[36] Of course, this is not to say that the 1960s drought or Hurricane Agnes were man-made, but rather to argue that those who see such calamities as emanating from objective, natural forces are likely to marginalize—if not demonize—any attempt to understand them with reference to human agency and morality. If these parables of wet and dry tell us anything, it's that the naturalness of a disaster is one part ecology to two parts ideology, and that one's ideology stems from where one stands on the technological domination of nature.

THE MORTON SALT DISASTER

When the natural-weather people singled out cloud seeding as having caused the Rapid City flood of 1972, they were on more solid ground than ever before in their stormy history.[37] Rapid City, South Dakota, is at the eastern edge of the Black Hills, in Mount Rushmore country. Rapid Creek, as the name suggests, is a white-water stream that plunges from the hills right through the city. In the late nineteenth century, settlers shied away from building homes too close to the stream for fear of flooding. But in the years after the Second World War, as the population of the city swelled, private homes and trailer parks squeezed their way onto the floodplain. Such development received a big boost from the Bureau of Reclamation in 1956, when the organization built Pactola Reservoir as a device for controlling floods. In 1972, some 43,000 people had carved out what must have seemed comfortable and safe lives for themselves beneath the bureau's water control project.[38]

Early in June 1972, a tropical air mass pushed its way across the western United States, causing flash flooding in the Mojave Desert that wound up burying cars under piles of mud. This powerful, huge air mass then proceeded north and east until it arrived in the Black Hills, precisely in time to clash with a stream of very cold air flowing down from the Arctic. The collision of the two air masses on June 9 produced a series of thunderstorms that refused to quit. Some areas in the Black Hills region received *15 inches* of rain in less than six hours. According to one government report, the rainfall west of Rapid City "averaged about four times the 6-hour amounts that are to be expected once every 100 years in the area."[39] Even worse, much of the rain fell on the roughly 50 square miles *below* the Pactola Reservoir. The bureau's flood control structure was thus about as useful as a foldup umbrella. A flash flood started, and when it ended 238 people had died, five of whom have yet to be found. Predictably, more than 500 mobile homes were totally mangled. That figure nearly equaled the number of permanent dwellings obliterated in the disaster (770), even though mobile homes composed just a small fraction of the total housing stock.[40]

Rapid City, 1972 (National Oceanic and Atmospheric Administration, Department of Commerce)

There is only one thing more incredible than all the rain and destruction: In the midst of the rainstorm, clouds were seeded as part of the Bureau of Reclamation's Cloud Catcher project. The bureau contracted with the Institute of Atmospheric Sciences at the South Dakota School of Mines and Technology in Rapid City to conduct the program. Two cloud-seeding missions were flown on the day of the disaster.[41] The first plane took off at 2:30 P.M. after detecting no immediate signs of severe weather. The plane dropped more than 300 pounds of Morton Salt into clouds located northwest of Rapid City. Before the launch of the second seeding mission, the men in charge of the project, Arnett Dennis and Alex Koscielski, discussed the advisability of another flight, given the chance of flooding in the hills area. By late afternoon, the wind had increased to 30 knots. They decided to send the plane up anyway, on a mission south of the city. The weather conditions during the flight were apparently so bad that the pilot had trouble controlling the aircraft. Before the plane had landed, Koscielski was on the phone to the National Weather Service, advising it of the possibility of flooding in the Sturgis area, precisely the area seeded earlier.

Pausing for a moment to reflect, we might say that such foolishness was exactly what the famous German philosopher Max Horkheimer had in mind in his 1947 book titled *Eclipse of Reason*. In it he wrote, "The more devices we invent for dominating nature, the more must we serve them if we are to survive."[42] And serve them we did that rainy day out in the Black Hills.

To Horkheimer's dictum might be added the corollary, that the more we dominate nature the more natural our disasters are made to seem. Soon after the flash flood, the governor of South Dakota, Richard Kneip, issued a press release assuring citizens that the cloud seeding had not contributed in any way to the destruction. According to the release, "Kneip said the last thing that is needed in an emergency such as Rapid City is going through is for unfounded fear or sensationalism concerning a scientific operation that scientists had reported had nothing whatsoever to do with the flood."[43]

The man the governor relied on for this assurance was Dr. Richard Schleusener, a one-time consultant to the Bureau of Reclamation and a major proponent of weather modification. Schleusener directed the Institute of Atmospheric Sciences in charge of Project Cloud Catcher. "I can assure you," he told the governor in no uncertain terms, "that the cloud seeding did not contribute to this disaster." It was, he declared, nothing short of "ridiculous to think that with a few hundred pounds of finely ground table salt disbursed from a single airplane we could cause twelve inches of rain in a few hours." Schleusener's conclusion was simple and emphatic: "There is no evidence that the 1972 flood was other than a natural event."[44] Arnett Dennis, who played a more direct role in the cloud seeding (and who, incidentally, had run the seeding program for the Blue Ridge fruit farmers before Wallace Howell took over the project), put an even finer point on the matter. He was dead certain that the seeding had nothing to do with the storm: "I would stake my life on that."[45]

Could the seeding experiment, designed to increase precipitation, have played no role whatsoever in the disaster? It was not an easy question, and it certainly required a much less categorical answer than the one served up by the staff at the Institute. Fearing a cover-up, Fred Decker, an atmospheric scientist at Oregon State University, wrote Sen. George McGovern (D-South Dakota), urging him to appoint a commission to examine the true causes of the calamity. "Human lives should not be snuffed out because scientists need the results of experiments," he wrote.

Decker also explained that he knew from experience "that some of our colleagues who are 'believers' about cloud seeding will hardly provide disinterested or objective advice in the wake of this tragedy."[46]

About that, he was right. Shortly after the calamity, none other than Wallace Howell, now an official with the Bureau of Reclamation in Denver, arrived in Rapid City to see for himself what had happened. He piled into a helicopter with Schleusener to observe the flood-ravaged city firsthand. Howell later met with state and local officials and told them that the seeded clouds were physically separate from the ones that produced the storm, implying of course that the disaster had occurred naturally. He also denied that Schleusener and his colleagues were asking for trouble when they decided to launch their cloud-seeding missions. In his opinion, the risks associated with weather modification remained small compared with the benefits to be gained, though of course those benefits had yet to be shown to exist, at least statistically speaking.[47] In all, what Howell had to say was what one would expect from a man who had dedicated the better part of his life to peddling weather modification.

Continued public attention, however, forced the governor to appoint an outside team to investigate the role of seeding in the disaster. Merlin Williams, the director of South Dakota's Weather Control Commission, chose the committee. South Dakota had long been a major center of cloud-seeding activity, with over one-third of the counties in the state experimenting with the new technology in 1951. Two decades later, the same year as the flood, South Dakota became America's first state to finance a cloud-seeding program, designed to boost precipitation and mitigate hail.[48] Obviously, were Project Cloud Catcher's seeding missions to be implicated in the Rapid City flood, the prospects for continued state support would be jeopardized. In other words, Williams had every incentive to organize a whitewash.

He chose three men to head up the investigation: Pierre St.-Amand, Robert Elliott, and Ray Jay Davis. St.-Amand worked at the Naval Weapons Center in China Lake, California, where he helped develop weather modification technology secretly used to fight the Vietnam War. The United States operated a top-secret rainmaking program between 1967 and 1972, designed to wash out roads and cause landslides along the Ho Chi Minh Trail. A major proponent of weather modification, St.-Amand later testified before Congress about his role in the scheme, saying that "the main thrust of my professional work has been oriented toward the safe and profitable use of the environment for human benefit."[49]

Then came Robert Elliott, the president of North American Weather Consultants, a private cloud-seeding firm based in California. For Elliott, one of the founders of the Weather Modification Association back in 1951, the Rapid City disaster must have seemed like déjà vu. In 1958, Elliott's firm wound up in court over cloud seeding it did for a California utility company, an operation linked to the devastating Yuba City flood that killed 37 people.[50] Though the cloud seeders eventually stood vindicated, the bad publicity generated by the case, Elliott noted, had caused insurance companies to grow queasy about weather modification, dampening enthusiasm for his services. Needless to say, weather modified disasters were bad for business.[51]

Ray Davis, a lawyer and the final member of the team, had carved out a reputation for himself defending commercial cloud seeders. Like St.-Amand and Elliott, Davis also played an active role in the Weather Modification Association, moving on to become its president in 1979. Davis and Elliott had, indeed, worked with Schleusener both to counter bad publicity and to boost the prospects of cloud seeding nationwide.[52] In short, a less impartial committee is hard to imagine.

The three-man panel began its work on June 21, 1972. Just one week later, by virtue of what Williams called "an almost herculean effort," the team released a draft report.[53] Not surprisingly, they concluded that cloud seeding did not contribute to the disaster. Floods, they said, occur naturally in Rapid City about once a decade. A 1907 flood had even swamped an area the same size as that affected in 1972. Their report was an excellent argument for clearing the floodplain of housing and letting nature regain control. And that, through federal government funding, is exactly what happened—one of the more intelligent responses to calamity on record in the United States (putting aside the fact that the investigators were supposed to be examining the role of cloud seeding in the disaster, not making recommendations on floodplain management). In any case, Williams could not have been more pleased with the committee's work: "I really feel that the whole field of weather modification has been strengthened by the report."[54]

The report did draw some criticism. H. Peter Metzger, a biochemist and writer, published a story in the *Denver Post* that called the committee's work a "scientific smokescreen." In truth, the double-talk it contained was enough to drive any logician to distraction. In the course of a single sentence, Metzger noted, the committee claimed both that the seeding did not increase the severity of the storm *and* that it was impossible to say what role the seeding played.[55] Metzger's article was later reprinted in the *Los Angeles Times*.[56] And inevitably, the *National Tattler* picked up the story. "GOVERNMENT WEATHER TAMPERING IS CAUSING WORLD FLOODS," a huge front-page headline blared. The report cited Metzger as well as an unidentified "independent weather expert." Said the mystery weatherman, "We keep telling these guys to cut it out (careless cloud-seeding programs). The only place we made any headway is with the Rapid City thing. They are scared to death."[57] The *Tattler* coverage prompted an immediate response from the *Rapid City Journal*, a longtime supporter of weather modification and a defender of Project Cloud Catcher. In an editorial titled "Hysteria Shouldn't Curtail Scientific Weather Modification," the paper croaked: "Far too many sensation-mongers are grabbing onto cloud seeding in relation to the June 9 disaster and merchandising fear and doubt in the worst tradition of yellow journalism."[58] Once again, the tendency to respond emotionally to disaster clashed with an attempt to restore normality—in this case the need for more weather modification to control nature's disorderly tendencies.

Perhaps the most reasoned scientific critique of the disaster came from Jack Reed, a meteorologist at the Sandia Laboratories in New Mexico. Reed published a short letter in the well-regarded *Bulletin of the American Meteorological Society* in 1973. Attempts to disassociate the cloud seeding from the flood were made in haste, he claimed. A look at the data "does not...necessarily lead a critical observer to this conclusion with the *overwhelming confidence demanded* where public safety is involved." One could in fact come up with a theoretical model, he continued, demonstrating a connection between the seeding missions and the calamity. Ultimately, "unacceptable risks *may* have resulted from the cloud seeding operations."[59]

In the end, the letter simply suggests the obvious: Cloud seeding could have contributed to the storm. Quite in contrast to Schleusener and the others involved in Project Cloud Catcher who dogmatically downplayed any connection between their work and the flood, Reed showed enormous restraint. Yet judging from the reaction of Schleusener and his colleagues, one would think that Reed had put forth outrageous accusations. In a letter to St.-Amand, Schleusener referred to the Reed "speculations" as contributing to the "folklore" surrounding the disaster.[60] He also tried to stop publication of the article. Merlin Williams even went so far as to lodge a complaint with Reed's employer, the Atomic Energy Commission, arguing that such antitechnological views were being made at "the tax payers' expense."[61]

Meanwhile, Robert Elliott placed Reed's work in historical perspective, noting that this was hardly the first time that cloud seeding had been singled out for spawning disaster. "Cloud seeders were most recently accused of causing the drought in northeast United States, and there are many residents of Pennsylvania, Maryland, and West Virginia who are still thoroughly convinced of this." The weather, he keenly observed, "seems to bear a 'speculative attraction.'" Before cloud seeding, "the atomic bomb was blamed for adverse weather events, and prior to that the introduction of telegraphy was blamed for bad harvest in England." While it is true that technical progress has proved fertile ground for the incubation of apocalyptic fears, it also bears mentioning that some technologies have a much tighter causal relationship with the catastrophe in question than others. And since a reasonably direct connection seems to exist between cloud seeding—designed, after all, to modify the weather— and atmospheric change, we can hardly fault the lay person for suspecting that the weather modifiers actually had caused calamities.[62]

Ultimately, the clash between the dominant technocratic culture and its subordinate counterpart turned on the question of the antidemocratic tendencies of applied science, as St.-Amand's response to Reed made abundantly clear. Although the public needed to be "protected from irresponsible activity," St.-Amand conceded, present laws were sufficient to accomplish that. "If every issue becomes a matter of local option—and it may so develop—then no further progress will be made in any branch of technology."[63] For St.-Amand, Schleusener, and the rest of the would-be controllers, genuine concerns for democracy and moral reason did not enter into their quest to determine weather in accordance with economic considerations.

CRANK

In the summer following the Rapid City disaster, someone bombed a trailer filled with cloud-seeding equipment in southern Colorado's San Luis Valley.[64] The paraphernalia belonged to Atmospherics Incorporated, a commercial weather modification company based in Fresno, California. Atmospherics, whose president, Thomas Henderson, traveled the same professional circles as both Schleusener and Williams, had been hired by a group of barley growers to suppress precipitation and especially hail during the crucial period in the summer when the crop is harvested. It is essential that barley not get wet during this time. As a major purchaser of grain from valley farmers, the Adolph Coors Company dreamed up the weather control program in the 1960s. Only ranchers and others in the drought-prone valley objected to the seeding, because they felt that it aggravated the region's dry conditions. In the summer of 1972, a meeting—required under a new law passed earlier in the year—took place to discuss whether Atmospherics should be granted a permit to seed clouds. Some 600 people attended the hearing, which lasted until 2 A.M. Most of those present objected to the seeding, though the permit was still granted, subverting what had promised to be a democratic process and perhaps explaining why had been driven to violence. Archie Kahan of the Bureau of Reclamation recalled loud applause at the hearing following religious arguments opposing weather modification as against God's will.

Religious opposition to cloud seeding was by no means uncommon. A study done in South Dakota in the early 1970s by two sociologists, Barbara Farhar and Julia Mewes, discovered that 40 percent of those interviewed held what the authors called a "religio-natural orientation." Asked whether they agreed with the statement that "Cloud seeding probably violates God's plans for man and the weather," two-fifths of those sampled nodded their assent.[65]

Schleusener not only disagreed with such people, he kept a file on them, product that he was of the Watergate era. He labeled the dossier, very simply, "Crank." Admittedly, the crank file does contain some real oddities. Francis Bosco of Lakewood, Colorado, who fashioned himself a "weather engineer" and a "practical meteorologist," tells of a plan to solve Denver's smog and unemployment problems all at once by burning some secret chemical and blowing it into the atmosphere.[66] It was not weather modification, per se, that Bosco objected to, but the failure of "the unethical group at Rapid City…to consult my advanced procedures."[67]

There are, however, materials in the crank file written by people with legitimate, honest religious and moral objections to weather control. Mrs. Adolph Hermann of Lemmon, South Dakota, wrote to complain about the seeding of a vicious hailstorm in 1968: "Seems it would be much better to let the Almighty God handle the weather situation."[68]

Gertrude Milton Walcher of Colorado Springs also wound up in Schleusener's filing cabinet. In a pamphlet titled *How to Kill a Beautiful World*, Walcher criticized cloud seeding and other attempts to, in her view, destroy the earth. "This thesis is being written because of the righteous indignation, anger and frustration of a subjective pioneer in a world obsessed by objectivity and materialism.... ," she begins. "It was a beautiful world when the laws of God and Nature were revered. Now, it seems that there is nothing that is sacred."[69]

Although Schleusener did not take Walcher seriously, we should. If nothing else, she offers the start of a useful critique of the weather modifiers. To these people, she points out, nature has no intrinsic meaning or moral value. "One would think that they would have some compunction about destroying God's handiwork, but evidently such an idea has no meaning to them."[70] For the cloud-seeding boosters, nature is little more than a resource, with no meaning outside its value as an instrument for human economic development. Moreover, those seeking to obstruct the domination of nature were seen by the seeders as premodern in their thinking. Opponents of weather modification were made out to be weak, superstitious, emotional, and too prone to accept acts of God or nature. But it is important to note that the natural-weather people were also threatening to Schleusener and his weather-modified world. The opinions of the so-called cranks showed that cloud seeding was not a normal or inevitable social phenomenon but something that weather entrepreneurs, government bureaucrats, and political leaders had dreamed up and foisted on unwilling sectors of the American public. Above all, the cranks were dangerous because they interfered with the attempt to naturalize natural disaster. When weather-modified calamities occurred, the seeders argued that larger natural forces drowned out the intended effects of the seeding. But the cranks had an uncanny way of taking the efficacy of weather modification seriously at precisely the wrong time. They caught the modifiers in the thick of their contradictions, showing them to be not only oblivious to the role of human responsibility and morality in disaster, but at times illogical, their common sense clouded by a technocratic dream of controlling the uncontrollable.

SEVEN

······················

Forecasting
at the Fair
Weather Service

If we can't control the weather, at least we can predict and warn people about it. For many, such advanced meteorological information is taken for granted as one of the great virtues of modern technological life. But not all Americans have benefited equally from the work of the National Weather Service (NWS), as events in Piedmont, Alabama, population 5,000, demonstrate quite clearly.

On March 27, 1994, Palm Sunday, over 100 parishioners of the Goshen United Methodist Church sat together in prayer when suddenly an F3 tornado tore the church to shreds.* Twenty members of the con-

---

* Tornado intensity is classified according to a system developed by the University of Chicago meteorologist T. Theodore Fujita. The Fujita scale begins at F0, signifying winds of 40 to 72 mph and "light damage," and proceeds sequentially to F5, indicating winds of 261 to 318 mph and "incredible damage." An F3 tornado denotes winds in the range of 158 to 206 mph and "severe damage." Historically, the vast majority of tornado-related fatalities have occurred as a result of F4 and F5 storms. The 1994 storm ultimately left 40 dead in Alabama and Georgia. Putting aside the Goshen church deaths, 75 percent of the other fatalities were among mobile home residents.

gregation lost their lives, including four-year-old Hannah Clem, the daughter of the church's minister, Rev. Kelly Clem. Some said it was an act of God. Piedmont's mayor, Vera Stewart, remarked that people were asking a lot of questions. "They said, 'Why did God do this? Was he mad because we were wrongdoing?' We have gotten a lot of negative letters, things like 'Where was your God while you were praying?'"[1]

About a month later, in an elevator somewhere in America: "Did you hear about those church people killed by that tornado?" an elderly fellow asked black nationalist Askia Muhammad. "Yes. It's too bad," Muhammad replied. "Did you notice that all the people killed were white?" the man asks. The elevator door opens. "It's payback. That's what it is. It's payback for what they've done to us," the fellow says before heading off. Muhammad later reflected on the exchange:

> For a moment, like my interlocutor—and I dare say many
> other black Americans—I entertained the thought (however
> crass and uncivil it may seem) that God might have been
> expressing, or might one day yet to come express, vengeance
> which is His and presumably His alone to express, against
> white Americans.[2]

The Rev. Kelly Clem, on the other hand, said that God had little to do with the tragedy. In the mind of this 33-year-old pastor, wind, not God, killed her daughter, though she did tell Vice President Al Gore, who toured the area immediately after the disaster, that a better system of storm warning might have prevented the calamity. Reflecting on the disaster two years later, Clem said, "There's nothing that we can do to protect us from death. We are human beings living on this earth, we are not immune to the wrath of nature."[3] Act of God or act of nature—what is notable about this dialogue on the calamity is the constrained understanding of moral responsibility that underpins it. In all the rhetoric that grew out of the disaster, the vast majority focused on the forces beyond human control—God or nature—responsible for the tragedy. Certainly nobody called it an act of man, or better yet, an act of man's inhumanity to man.

Yet Kelly Clem had been correct in her initial admonition to Gore. Hannah Clem and the others might be alive today had the area been serviced by the federal government's weather radio system and outfitted with sirens (though, as we will later see, that system is far from consistently effective). At the time of the disaster, the National Oceanic and Atmospheric Administration's Weather Radio (NWR), which provides instantaneous notification of weather alerts, covered roughly three-fourths of the country. Piedmont, however, like many other rural areas, sat beyond the 40-mile reach of the nearest transmitter. It was not just a question of money issues delaying the complete implementation of the weather radio program. Increased personnel were needed to operate and maintain the service, a problem for the chronically short-staffed NWS. As far as sirens were concerned, the Calhoun County Emergency Management Agency sought funding for them in Piedmont but came up empty-handed—a common problem in poor, rural areas.[4]

After the disaster, Gore promised to upgrade NWR to cover 95 percent of the nation. But a year later, as Piedmont braced itself for yet another season of tornadoes, the warning system and sirens were still not in place. "Somewhere, somebody got bogged down," explained Mayor Stewart in a generous moment.[5] By the summer of 1995, the transmitting tower for servicing the Piedmont area with NWR had been erected and federal money set aside for two warning sirens. But still private funds had to be raised in order to pay for five additional horns needed for adequately warning citizens.[6] By the end of 1995, thanks in large part to the Clems and others dedicated to safeguarding the town, the warning system was in place. "It is somewhat unfortunate that we make progress in this country on the blood of victims," said Dr. Elbert "Joe" Friday, then director of the NWS. Such a statement implied, of course, that something could have been done to avert the disaster, a point confirmed by Friday's superior, Secretary of Commerce Ron Brown. "That tragedy, as we all know, need not have happened," he said—stunning words coming from the head of the department that has, in league with other arms of the government, done just about everything in its power to chisel Americans out of adequate weather services.[7]

As the sad unfolding of events in Alabama suggests, natural disasters do not always just happen; they are often produced through a chain of human and natural occurrences. And yet, the tendency to see nature or God as the primary culprit in tornado and other severe-weather calamities persists, supporting the massive commitment of technological resources that fail to address the underlying social causes of weather-related catastrophes. Under this scenario, disorderly nature—as opposed to rampaging social and economic forces—is seen as the problem that technology must solve. But natural disasters are not simply scientific dilemmas in need of a technical solution. They are instead the product of particular social and political environments. The technology existed for staving off the nation's post-1970 severe-weather disasters. What did not exist was the political will to properly staff the deployment of these forecasting tools. In this sense, one can point to any number of tragedies, apart from Piedmont, that could have been avoided—and the 1976 Big Thompson flood, the 1977 Johnstown flood, and the 1984 North and South Carolina tornado storm come to mind.

THE "T" WORD

The United States is a prisoner of its own geography. It occupies a spot on the planet at the intersection of two dominant streams of air: warm, moist air that flows north from the Caribbean and the Gulf of Mexico and dry, polar air that drifts south from the Arctic. When these two air masses collide the result is often severe weather—thunderstorms, flash floods, and tornadoes. No nation on earth has more extreme weather phenomena than the United States.

It is thus something of a surprise to learn that before the early 1950s, official tornado forecasts were not issued by the U.S. Weather Bureau. It was not because no one had thought of the idea. As far back as the 1880s, John Finley of the Signal Service, charged by Congress in 1870 with overseeing the nation's weather service program, developed a set of techniques for predicting tornadoes.[8] But the Signal Service, evidently somewhat unimpressed with Finley's forecasting powers, banned the use of the word *tornado* in any officially released forecast in 1883. The ban was removed in 1886 but then reinstated the following year. When the Weather Bureau emerged in 1891 to take over from the Signal Service as the nation's official weather agency, it, too, prohibited the word in forecasts. Indeed, the ban remained in place until 1938.[9] As late as 1943, explained historian Donald Whitnah, the Weather Bureau felt that tornado alerts "were more trouble than they were worth."[10]

That the word *tornado* was eschewed in weather service circles through the 1940s is clear. The logic, however, behind this act of self-censorship is somewhat obscure. In part, a concern with credibility must have driven this decision. After all, forecasts can easily turn out to be wrong, something even more likely in the period before radar came into use in the 1950s. Also, needlessly alarming people made them less likely to credit future warnings, a problem plaguing the NWS to this day.

Ultimately, the military took the lead in introducing Americans to the world of tornado forecasts. After a twister dropped out of the sky and destroyed 32 airplanes at Tinker Air Force Base outside of Oklahoma City in 1948, the military called on two of its best weathermen, Ernest Fawbush and Robert Miller, to provide advanced warning for severe weather. The forecasters succeeded in providing a tornado watch for the base just five days later, when another storm racked the installation. More important than their actual prediction was their pioneering work in creating the so-called tornado watch-box, a large area (somewhere on the order of 25,000 square miles, or about five times the size of the boxy state of Connecticut) singled out as likely to experience a tornado within a designated period of time. Word soon leaked out about the military's successful forecasting venture, and in Oklahoma the Weather Bureau came under fire from the press for not providing the service. Bad press in combination with pressure from midwestern members of Congress eventually led the Weather Bureau in 1952 to issue tornado forecasts.[11] Those forecasts became increasingly accurate in the late 1950s, when the agency installed radar in its offices.

Informing the public about such severe weather forecasts was of course another matter. The 1950s saw a huge increase in the use of television, which aided the dissemination of tornado forecasts directly into households. Late the following decade, the Weather Bureau launched its NWR program, erecting transmitters that allowed anyone with a receiver to learn of a posted tornado watch. The NWR system actually began in the 1950s, but not until the advent of affordable radio receivers was any significant headway made.[12] The greatest impetus for weather radio, however, came on the heels of disaster itself. On April 11, 1965, Palm Sunday, 50 tornadoes—including 19 of the most violent F4 and F5 classification—raked the Midwest, killing 256 people. In Goshen, Indiana, Sheriff Woody Caton reached a mobile home park shortly after a vortex leveled 92 trailers. "It looked like a giant auto-crushing machine had simply chewed the place up."[13] According to one Weather Bureau official, "slow public reaction to tornado warnings was the biggest contributor by far" to the high number of fatalities in the storms.[14] In response to the disaster, 77 weather radio transmitters went in over the course of the next five years—hardly a forceful response in a nation where three-fourths of the world's twisters touch down.[15]

Twin funnels bearing down on Elkhart, Indiana, April 11, 1965 (National Oceanic and Atmospheric Administration, Department of Commerce)

BLOOD AND PROGRESS

Thus we must take issue with Dr. Friday's correlation between blood and progress. If evidence from the 1970s is any indication, an inverse relationship may well exist between deaths in tornadoes and movement toward a more effective warning system. Consider the 1974 outbreak, when a stunning 148 twisters (30 classified as F4 or F5) laid waste to the stretch from Decatur, Alabama, all the way to Windsor, Ontario, killing 315 people. It was the fourth deadliest tornado event of the twentieth century, and the worst since 1932. Ten states were declared disaster areas, and property losses were estimated at $600 million, making it the most costly tornado outbreak in U.S. history.[16] If we want to understand the historical roots of the 1994 Piedmont calamity, the tornadoes of two decades earlier are a good place to begin.

Immediately after the disaster, the National Oceanic and Atmospheric Administration (NOAA) sent out a team of officials to investigate, as it routinely does after major calamities, in the hope of warding off a repeat performance. The team's findings were unambiguous: "Expand NOAA Weather Radio as quickly as possible."[17] Given the magnitude of the tragedy and the strong advice to expand weather radio, one would think that the government would have moved quickly. It didn't. Two months after the storm, when it became clear that congressional action would not be forthcoming, a reporter asked an aide to Sen. Thomas Eagleton (D-Missouri), who himself favored increased support for warning services, why Congress remained so unmoved. "Congress views the tornado," he said, "as an act of God where even the Congress can't intervene."[18]

Even two years after the deadly 1974 tornado outbreak, the weather radio system had received disgracefully little attention. The country stood little better prepared for another tornado outbreak, with still only 70-odd transmitters and major gaps in the Midwest and South. Although a focused effort could have resulted in 90 percent coverage within two and a half years, NOAA's plan called for total completion of the system in twice that amount of time.[19] Meanwhile, poor rural people, many of them in the South, were going to pay the price for the delay. Allen Pearson of the National Severe Storms Forecast Center in Kansas City explained that many of the areas most at risk for tornado disaster were economically depressed parts of the South, where the poor often lived in mobile homes. Commenting on the political economy of risk responsible for putting rural southerners in danger, Pearson explained, "Frankly, the quality of the warning system a community has in this country often depends on money, and the per capita income is higher in the North."[20]

It took money to install sirens and transmitters—this much is true. But it required more than just dollars to implement the national weather radio system. To operate NWR properly demanded a substantial investment in time and labor, and it may well have been this obstacle, more than any other, that led to the delay in its implementation. Thorough coverage consisting of a network of 331 transmitters would have required the production of over 10,000 scripts every day throughout the nation—

a gargantuan workload for the short-staffed weather service.[21] Labor problems, as we will see in a moment, have long been a problem at the NWS. John Townsend of NOAA summed up the 1974 tornado outbreak by observing that such situations stretched the nation's forecast and warning system thin. "We are programmed for average weather conditions."[22] In other words, the NWS was staffed for fair weather, even though in a typical year the United States was—and still is—subject to some 10,000 severe thunderstorms, 5,000 floods, and 1,000 tornadoes.[23]

MISSING TEETH

No set of disasters better demonstrated the weaknesses of the NWS's warning and preparedness operations than the three devastating flash floods of the 1970s at Rapid City (S.Dak.), Big Thompson Canyon (Colo.), and Johnstown (Pa.). As already noted, 238 people lost their lives in the Rapid City flood of 1972. Flash flooding, however, continued to cause problems throughout the decade, as the effects of development and the creation of more land area impervious to water caused runoff during rainstorms to increase in speed and volume. More people also came into contact with such floods, as remote canyons proved more attractive to tourists in search of wilderness adventure.

It is one thing to have rain on your vacation, but what happened beginning on the night of July 31, 1976, in Big Thompson Canyon—a steep 25-mile stretch along the river of the same name, about 40 miles north of Denver—went well beyond inconvenience and into the realm of nightmare. Some areas of the Big Thompson watershed experienced 12 inches of rain in the space of just a few hours. Trailers were sent careening downstream while huge propane tanks, "spinning like crazy," as one officer on the scene put it, began exploding. The force of the water so mangled the bodies of the 139 victims claimed by the flood that five dentists and eight FBI fingerprint experts had to be called in to make identifications.[24]

The following year, Johnstown—the site of the 1889 storm and flood that swept over 2,000 to their deaths—witnessed tragedy yet again along the banks of the Conemaugh River. Starting on the evening of July 19, it rained for nine hours, and hard. During one 40-minute stretch alone it rained over two inches. Johnstown's mayor, Herbert Pfuhl, described the ensuing flood this way: "It was like walking to the top of a mountain and coming face to face with a 15-foot high wall of rampaging water that was going to wash you back down." Seventy-six people died in the hideous deluge.[25]

Taken together, the three floods highlighted one main weakness of the NWS's response to disaster: a chronic shortage of staff. The staffing problem, the floods made clear, severely affected the NWS's ability to maintain its community awareness programs, an integral link in disaster preparedness. At the weather service office in Rapid City, a vacancy on the staff required the meteorologist in charge of the office and his principal assistant to perform more shift work, that is, engage more in the everyday responsibilities of forecasting. The increased shifts cut into the time they could spend developing an effective flood warning system in the community.[26] Similarly, in the Denver office responsible for the Big Thompson region, no disaster preparedness position existed, helping explain why police—dispatched to the canyon to warn people of danger—had such a hard time evacuating the area. "They looked at me like I was crazy, most of them," said one patrolman about those who refused to move to high ground.[27] As for Johnstown, there was no disaster preparedness specialist at the Pittsburgh office. Worse yet, illness combined with training assignments had prevented the staff from visiting the devastated area in the two years before the flood, leaving people scandalously unprepared.[28]

The sources of the NWS's inadequacies in the area of disaster response are obvious to anyone who can add a column of figures. At the time of the Johnstown flood in 1977, there were 5,015 full-time employees at the NWS. Ten years earlier, 5,022 people worked there. The NWS's budget in 1977 was $92.6 million (adjusted to reflect 1967 dollars) versus $91.6 million a decade earlier.[29] Imagine trying to run a complex forecasting network involving hundreds of offices with 1967 resources in a 1977 world, when the population of the country had increased 9 percent over the decade—with people crammed onto ever more marginal land—and the number of severe weather events showing no sign of abating.

The staffing figures especially are even more outrageous in light of a study done in 1969 at the request of the Department of Commerce—the very agency housing the NWS—indicating the need for 840 more employees. That represented an increase of 17 percent, an increase so large it was, of course, destined never to happen. The report called for 214 more personnel above the 1969 figures to bring the field office staff to what the writers of the report called "basic minimum levels."[30]

By the late 1970s, the number of fully staffed surface weather stations had declined by roughly 10 to 15 percent from two decades earlier. The grim situation prompted Allen Pearson, the director of the National Severe Storms Forecast Center—a man known affectionately to many as Mr. Tornado because of his obsession with violent weather—to observe:

> There are times at night, for example, in the State of Nebraska, which is a fairly good-sized State, that we have only two or three stations open and you could bury a tornado or even a small hurricane in some parts of Nebraska at night and if you depended upon surface stations to alert you, you couldn't find it. It's like taking a comb and you break off one tooth today and then next couple of days you break off another tooth and pretty soon all of a sudden you can't comb your hair any more.[31]

If the weather service's problems were not bad enough, in 1978 budget constraints forced the closing of 19 offices across the nation. The move further limited the amount of data available in the event of severe weather. Dr. Robert White, who headed the Weather Bureau in the 1960s, objected to the shutdowns. He told Congress that the nation's network of observation stations had already been stretched too thin: "I am concerned that a further reduction in the density of those stations will provide us with situations in which we do not have adequate information to define smallscale storms." Once again, the personnel shortage remained at the root of the NWS's problems. To conserve labor, the weather service constantly felt pressured to close more offices. "Time after time in weather disaster situations we have seen manpower stretched dangerously thin...," White noted. "I hope it will not require some tragedy to force the realization that manpower is indeed insufficient."[32] These words came in 1978, and we can only surmise that a body count of more than 450—the combined death toll in the three major flash floods that decade—was not disaster enough to force action. Yet another instance where more blood has not spelled more progress.

READY...GO!

In 1985, four fishermen out in the Atlantic Ocean were killed in a storm. Their families turned around and sued the National Weather Service for negligence and won over $1 million.[33] That judgment, awarded by U.S.

District Judge Joseph Tauro, turned out to be one of President Ronald Reagan's favorite rallying cries for tort reform. *Imagine* holding the NWS responsible for blowing a forecast, the lament ran. The next thing we know people all over America would be suing the government all because a bunch of weathermen made a mistake.

But inquiring into the deaths of those fishermen reveals that Judge Tauro's decision is not as far-fetched as some would allow. On November 21, 1980, two ships, the *Fairwind* and the *Sea Fever*, made their way from Hyannis, Massachusetts, toward Georges Bank to trap lobsters. Both ships were properly outfitted with all the requisite nautical equipment and both closely monitored the weather forecasts. Buoy data on wind speed and sea heights in offshore locations was among the NWS information regularly consulted on such trips.

Before setting off to fish, the captains heard a favorable forecast. As they sailed for Georges Bank they tuned in to hear yet another indication that fair weather awaited them at their ultimate destination. But by the time the ships made the final leg of their journey, arriving about 11 P.M., the weather had worsened; yet the official NWS forecast at that time still indicated good weather for fishing. Not until nearly 5 A.M. the following morning did the weather service issue a warning about inclement weather. By this time the winds were already at gale force, with 25-foot seas, according to Capt. Peter Brown of the *Sea Fever*. Things only got worse. By 11 A.M. the winds reached 70 to 80 knots and the seas had swollen to 30 to 50 feet. The *Fairwind* capsized, throwing its captain and three crew members overboard. Miraculously, crewman Ernest Hazard survived for two days on a raft in nothing but underwear before the U.S. Coast Guard rescued him. The other three men perished. Although the *Sea Fever* fared better and did not sink, one crew member fell overboard and died in the roiling waters of the North Atlantic.

Crew members thought that the NWS had based its forecast on information provided by a buoy located at Georges Bank. In fact, the buoy remained out of service. The buoy's wind sensor had failed entirely almost three months before the storm, but the NOAA Data Buoy Center, in charge of the contracting out of buoy maintenance and repair, did not replace it. The buoy, though inoperative, stayed on station.

The families of the dead seamen sued NOAA and the NWS for negligence in not repairing the buoy. Judge Tauro sided with the victims. The buoy, he opined, was absolutely critical to the Georges Bank forecast; the

NWS was aware of this, and yet let the broken device languish. Nor did the NWS ever inform the lobstermen that its forecasts were based on incomplete information. Put simply, the law says that the government does not have to provide weather forecasts for mariners. But if it chooses to do so, by law it must not be negligent in carrying out its duty. Tauro claimed NOAA had acted negligently by failing to fix a device it knew had broken.[34] The First Circuit Court of Appeals, however, disagreed and overturned the decision.[35] If the NWS had knowingly used faulty data—something of course that only those with the most shameful disregard for human life would do—then it would have been negligent, said the higher court. The appeals court applied a narrow understanding of legal liability and thus the government escaped. Yet Tauro's decision was not the least bit strained. It simply reflected the application of a broader—but by no means wrong—set of principles for legal liability.

In its opinion, the first circuit expressed concern that by ruling for the plaintiffs the court would be opening the floodgates for all kinds of suits against the U.S. government. In this respect, one can perhaps understand their logic. But that said, the decision still helped to paper over the human economic forces involved in the tragedy. We might, for example, question whether the buoy would have gone unrepaired had the government itself been directly responsible for fixing the problem instead of contracting out the maintenance and repair to a private company.

In the mid-1970s, James Winchester, who ran the data buoy office, elevated the contracting out of services at NOAA into a kind of religion. Winchester was a privatization fiend who during the Reagan years went on to rise within the ranks of the agency. Needless to say, his ascent spelled more trouble for the already beleaguered NWS.[36]

To muster support for his vision of less government involvement, Winchester, an associate administrator at NOAA by the early 1980s, hired the consulting firm of Booz-Allen and Hamilton to study the structure and organization of the NWS. Booz-Allen's answer to the weather service's problems was simple: finish dismantling the entire operation. Its report recommended that the field structure of more than 300 offices be reduced to a mere 50 or less. Booz-Allen, in other words, planned to do away with the backbone of the NWS's warning system. The existing weather service forecast offices, responsible for issuing state forecasts (of which there were 52) would be kept at their current level or cut in half. Forget about warning services, Booz-Allen seemed to be saying. Forget about training spotter networks critical for tracking severe weather. Forget about developing close ties between the NWS and the public most at risk for severe weather. Jettison weather radio, which by this point had finally been deployed to a respectable 367 stations, and let privately employed forecasters issue weather service products.[37]

The Booz-Allen study amounted to a neat little handbook that should have been titled: "How to Have a Natural Disaster Without Really Trying." It flew in the face of all the reputable scientific studies done on the NWS, as well as the lessons taught by the major disasters of the 1970s. The year before Booz-Allen issued its study, the National Advisory Committee on Oceans and Atmosphere made its own report on the weather service. Its conclusion could not have been more different. "We are concerned about the weakening of the service that can result from increased closure of weather stations," the advisory committee stated. "Evidence indicates that the effectiveness of weather services are greatly enhanced when there is a close relationship between communities and local weather services that serve them." The committee, chaired by John Knauss, a highly regarded oceanographer, expressed concern that continued personnel and resource shortages were hobbling the agency. "We believe that the National Weather Service is now at a point where it may not be able to provide the warnings and protection expected of it in the face of major natural catastrophes."[38] Keep the stations open, the report intones. Hire more staff. Invest more money and resources in the nation's warning system. Develop the interface between the weather service and the people likely to experience bad weather—the weak link in the warning chain, as pointed out by all studies.[39]

In 1984, there came a disaster that offered a glimpse of the weather service world according to Winchester and Booz-Allen. Some 20 tornadoes swept through the Carolinas, injuring 1,300 and killing 59, making it the most deadly outbreak since 1974. At least five of the tornadoes rated F4, indicating "devastating" damage from winds that may have ranged as high as 260 mph. Predictably, nearly 40 percent of those killed, including a pregnant woman and her mother from the small town of McColl, South Carolina, perished in mobile homes. The destruction was so thorough, the scenes so positively macabre—replete with cows found slumping from trees—that people turned to the hyperrealistic world of television to describe the aftermath. "The only thing I've ever seen like it is that film, 'The Day After,'" said Mayor Johnny Weaver of Bennettsville,

South Carolina, referring to the made-for-TV movie about a nuclear holocaust that annihilates Kansas City, Missouri.[40]

"It looked like the hand of God reaching down to crush us," said Joyce Leonard of twister-wracked Mount Olive, North Carolina.[41] But in truth, more than one hand was at work here. Although the weather service produced an accurate tornado watch for the affected counties in both North and South Carolina, it failed to issue ample tornado *warnings*. Watches indicate that conditions are ripe for a tornado; warnings mean that someone—often a spotter who volunteers to look for funnels—has actually seen one. The lack of warnings meant little lead time to run for cover; hence the 59 deaths. After the storm, a NOAA disaster survey team discovered that tornado spotter networks in both states were either absent or of limited value. The reason was simple: The weather service offices located in the affected areas simply did not have the staff to develop adequate warning preparedness networks. South Carolina had a warning preparedness meteorologist on staff at the Columbia Weather Service Forecast Office. But she spent less than 50 percent of her time promoting community preparedness, because a staff shortage pressed her into doing routine forecasting shifts. North Carolina did not even have a warning preparedness meteorologist.[42]

In congressional hearings held later, Andy Park, who headed the weather department of a North Carolina television station, questioned the efficacy of the NWS's entire watch and warning program. Park, who claimed to be "just a poor old country boy that's spent 25 years studying the weather," wondered whether issuing a two-stage alert gave people enough lead time to get out of harm's way.[43] It was the right question to be asking. Indeed, a report done by the National Academy of Sciences as far back as 1977 had urged the NWS to consider a three-stage warning system. The panel suggested inserting an intermediate advisory between the watch and the warning—perhaps calling it an "alert"—and issuing it one to two hours before a severe weather event. "The hierarchy of three messages—watch, alert, and warning—should be easily understood in the context of 'ready, set, go,'" the panel wrote.[44] Under the new system, advisory statements would be more precise, giving people more notice about an impending calamity. The innovation was never adopted, however. Harried forecasters, already suffering from the effects of chronic staffing shortages, would simply not have been able to accomplish the added workload of issuing another statement. Ready...Go!—it would have to be.

THE FORECAST FROM BERLIN

It must have seemed eerily familiar four years later when an F4 tornado beat an 83-mile track through North Carolina. Damages amounted to $77 million and almost 1,000 people were left homeless. Leroy and Mary Alston of Nash County went to sleep and probably never knew what hit them. A funnel cloud picked up their mobile home in the early hours of the morning and smacked it against a tree. The couple was later found dead, still locked in each other's arms.[45]

"Miraculously," says a NOAA investigative report, "only four people died" (though 154 were injured). *Miraculous* was the right word. Again, chronic staffing problems that for some 15 or more years had plagued the weather service interfered with its ability to protect the public. To its credit, the weather service did prevail in appointing a warning prepared-ness meteorologist at the Raleigh office following the 1984 disaster. If only proper staffing had allowed him to do his job, preparing people and saving lives. In fact, because of staffing vacancies, he had to cut back on community preparedness. Each year, as the need for shift work asserted itself, he worked less and less at preparing people for a natural disaster, devoting 62 percent of his time to such activities in 1985, 58 percent in 1986, 52 percent in 1987, and 48 percent in 1988. Other staffing problems contributed more directly to the 1988 tornado disaster. The Wilmington weather office issued no warnings about the tornadoes, leaving people completely unprepared. In large part, the presence of only one over-worked meteorological technician on duty at the time of the storm led to the oversight. Normally, only one person covered the midnight shift. But during severe weather that person would generally be expected to ask for help. In this case, however, notes NOAA's report on the calamity, "the man on shift decided to 'tough it out,' partly because of the reduced number of people available to work at WSO Wilmington NC, and partly because of the known concern to keep overtime costs down."[46] Again, as had happened in 1984, the weather service found itself understaffed and overwhelmed.

There was, however, one difference between the 1984 and 1988 calamities. In the latter case, a mechanical problem helped contribute to the disaster. It was about the most serious problem you can have in the weather forecasting business: The radar was out of order. During the entire week preceding the storm, Raleigh's radar remained inoperative as the office sat waiting for parts. By the time the right parts finally arrived, the storm had already blown through. "We're operating in the Paleolithic age in our country in terms of weather forecasting," said Congressman James Scheuer (D-New York).[47]

What is most striking about Scheuer's comment is that he made it in 1990, after a decade spent modernizing the NWS's systems. Modernization of the weather service began in 1980, when the agency signed a joint agreement with the Federal Aviation Administration and the Department of Defense to develop something called NEXRAD, short for Next Generation Weather Radar. The existing radar network had been deployed during the 1960s and 1970s, though much of it was based on 1950s technology. NEXRAD, in contrast, consisted of the latest Doppler radar technology, as well as computers and communications equipment. The new Doppler-based radar excelled over the conventional kind in a number of respects. Conventional radar sends out a beam that then reflects off, say, raindrops, allowing a weather forecaster to detect the magnitude and location of a storm. Doppler-based radar, however, works by using microwave pulses to detect and measure movement. Thus a weather fore-caster using Doppler technology can, by using a computer to calculate the shift in frequency of the pulses as they return, tell the direction and speed of a storm system. In addition, while conventional radar can provide a vague idea of storm intensity, Doppler technology can measure the wind directly. Doppler thus held out the promise of increasing the lead times for tornado warnings, something that can make all the difference when you are diving for cover.[48]

To build the NEXRAD system, the government called on the Sperry/Unisys Corporation, the defense contractor. Apparently, the company had a little help from James Scheuer—no Paleolithic man he—in securing the business. It seems that at the same time Scheuer was holding hearings in the mid-1980s on the selection of a radar contractor, he was up to his elbows in Sperry/Unisys stock. "I am flabbergasted to hear that I was buying—or that somebody on my behalf—was buying Sperry stock for me at that point in time," he told NBC's *Dateline* in 1992.[49]

Modernizing the weather service with NEXRAD and other new technologies was of course expensive, so the question is raised: Why after years of slowly bleeding the NWS would the Office of Management and Budget (OMB) agree to finance a costly new modernization program? Richard Hirn, the lawyer for the National Weather Service Employees

Organization, believes that a Faustian deal may have been made between the weather service and OMB. In return for the money to fund its modernization program, the weather service agreed to a drastic reduction in offices.[50]

The existing office structure had evolved over the course of the late 1960s and early 1970s. In order to meet specific community needs, offices were located in coastal areas prone to hurricanes and in places such as Harrisburg, Pennsylvania, where a sizable population must contend with river flooding. Under the old field office system, 52 Weather Service Forecast Offices produced state forecasts, and 250 Weather Service Offices (WSOs) made weather observations, issued severe weather warnings, and engaged in community interaction.[51] Under the new modernized weather service plan, however, the office structure was simplified and reduced. As originally conceived, the backbone of the new system would be 115 Weather Forecast Offices (WFOs). All the WSOs would disappear. In other words, the weather service would go from operating over 300 offices to just 115. The WFOs would be neatly spread out across the map, with each office responsible for forecasting the weather within a radius of roughly 125 miles, the supposed reach of the NEXRAD units. When it came to office location, questions of service to the community took a back seat to arbitrary technological imperatives. Modernization centered on standardizing the offices and spreading them out as equally as possible in order to maximize coverage. But some people worried that closing down the local weather offices would lead to a collapse of spotter networks, not to mention the entire warning preparedness effort. Said Wayne Broome, emergency management director in Mecklenburg County, North Carolina, in 1989, "When I heard what they planned to do, it was almost like somebody told me I had cancer."[52]

There was of course nothing inevitable about this new meaner and leaner office structure. The NWS could have deployed the NEXRAD system and kept the old complement of offices. Indeed, a plan recommended by a Department of Commerce and NOAA management team in 1984 suggested just that.[53] The impetus behind retaining all 300-odd offices was the long-standing belief that a close physical and working relationship between NWS personnel and local communities enhances the quality of weather services and saves lives.[54] Moreover, if the weather service was interested in saving money instead of closing offices, it could

have foregone NEXRAD and settled for less technologically sophisti-
cated but still valuable Doppler units built, say, by Enterprise Electronics
in Alabama, a company that local television stations and research uni-
versities turned to for their radar needs.[55]

Although costly in terms of the money and time it took to develop,
NEXRAD is unquestionably an extremely valuable tool for weather forecasting. "The capability of these new radars exceeds, in practically every
respect, the radars that they replace," explains an important recent study
showing that NEXRAD improved warning lead times.[56] But like any
technology, it has its limits. While it does an excellent job picking up
severe thunderstorms and intense tornadoes, it has trouble detecting
weaker intensity tornadoes, common in the Carolinas. Also, Doppler
radar is most effective at detecting tornadoes that are approximately 30
miles away, its ability decreasing in a linear fashion to zero at 100 miles.[57]
Yet the 115 planned NEXRAD-equipped offices were commissioned
based on the assumption that each one would have a 125-mile radius of
responsibility, a radius that—from the standpoint of tornadoes—the
Doppler technology at least would not be able to cover.[58] Although some
members of the NWS's management conceded NEXRAD's limitations,
many persisted in glorifying and, at times, overselling the new technology. Joe Friday called NEXRAD "the most significant improvement in
weather warnings in my 32 years at the National Weather Service." Said
Ron Alberty, director of the NEXRAD support system in Norman,
Oklahoma, "I would stand on my reputation that when the network is
complete, no major thunderstorm or killer tornado will go undetected."[59]
Don Witten, a weather service public affairs officer, said it would provide
"an umbrella over the United States."[60] Even if they believed these words,
such statements created a false sense of security that the new technology,
whatever its merits, did not completely warrant.[61]

Implicit in the entire NEXRAD-driven modernization program was
the assumption that the new technology could be used to annihilate
space, not to mention offices. Meteorological wisdom had long held that
forecasters make their best forecasts in the local area. But the moderniz-
ers felt that no matter the location of a weather forecast office, all areas
within the 125-mile radius could expect a similar level of service.
Nowhere was this clash between modernization and professional meteorological wisdom more evident than in the NWS's attempts to relocate
certain offices. Perhaps the greatest public outcry involved the moving of

The National Weather Service's view of NEXRAD's near total national coverage (Reproduced from National Oceanic and Atmospheric Administration, *Strategic Plan for the Modernization and Associated Restructuring of the National Weather Service* [1989])

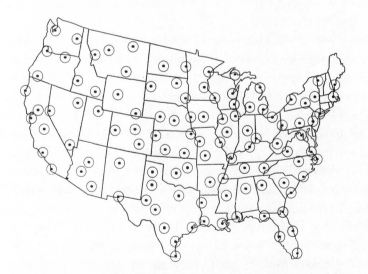

A less optimistic view of NEXRAD provided by the National Weather Service Employees Organization (Reproduced from National Weather Service Employees Organization, *Response to the Strategic Plan for the Modernization and Associated Restructuring of the National Weather Service and Recommendations* [1989])

the Redwood City office in the San Francisco Bay Area to Monterey, approximately 100 miles to the south. Ground was broken for the new office in early 1993. At the time, Monterey was reeling from the closure of the huge Fort Ord military base and, while a new weather service office would hardly generate the economic returns of a vast military installation, it represented a positive development for the area nonetheless.[62] There were also other economic reasons favoring the Monterey site, such as the 8 percent cost-of-living differential—paid by the federal government to employees living and working in San Francisco—to be saved in the move.

The move to Monterey generated enormous opposition, mainly because the relocation seemed to go against meteorological common sense. John Monteverdi, chairman of the geosciences department at San Francisco State University said, "On a scientific level, I'm kind of bamboozled by this move." He believed the decision would degrade forecasts for the six million people living in the San Francisco Bay Area. "The move is directly counter to both good meteorology and conventional wisdom," said Jan Null, a veteran forecaster at the Redwood City office. "There's no question that the forecasts aren't going to be as accurate."

To be sure, a great deal of resistance to the move among the Redwood City forecasters occurred because relocation threatened to disrupt their personal lives. But the forecasters were not the only ones opposed to the move. Dan Fowler, an emergency services consultant in the San Francisco area, called the decision to relocate "a stupid move."[63] William Medigovich, regional director of FEMA, noted that the Redwood City office offered major support through its detailed forecasts during the devastating Oakland–Berkeley Hills fires of 1991 and feared the loss of face-to-face meetings with weather service personnel.[64] "It's totally frightening and scary, and it's criminal," remarked Steve Johnson of the Central California Weather Observer Association. "They're going to kill somebody."[65] Whatever the impact on human life, the idea that new technology could be used to forecast equally well and provide the same level of weather service to all areas within the 125-mile radius was wishful thinking. Exclaimed the National Weather Service Employee Organization (NWSEO), another opponent of the move, "If one can argue that a forecast office in Monterey can serve the San Francisco Bay area well, then one could also argue that the NWS should build a huge forecast center in a bunker in Kansas and forecast from there for the entire nation!"[66]

Better yet, why not move the entire forecast operation to another country? The weather service modernization plan called for closing two offices in Harrisburg: the WSO and the Middle Atlantic River Forecast Center (RFC) located two blocks from the Susquehanna River, one of 12 such centers nationwide for predicting floods. The two offices were to be moved northwest to State College, Pennsylvania. With the relocation, the river forecast center would, of course, lose its tie to the river, a connection that had come in handy during the flooding caused by Hurricane Agnes in 1972. The storm knocked out a vital communications network, obstructing efforts to collect the data necessary to forecast the cresting river. But being in Harrisburg on the eleventh floor of a building, RFC staff were able to use binoculars to evaluate flood heights. It was also useful during the winter for RFC staff to be able to see the Susquehanna in order to forecast flooding sometimes caused by ice jams.[67] So why did the weather offices move from the well-populated Harrisburg area on the Susquehanna to the less dense site near State College? According to Herb White, director of public affairs for the NWSEO in 1992, the only reason the NWS gave was the desire to be near Penn State and its well-regarded meteorology program. White suspected that two members of the university, Charles Hosler, of weather modification fame, and John Dutton, both of whom played key roles in overseeing the weather service's modernization plan, may have used their influence to persuade management to relocate. "Charles Hosler and Dr. Dutton are very prominent members of the meteorological community with contacts," said White. But relocating a river forecasting center off of a major river, where it could best visually monitor water levels and the formation of ice, seemed more than a little counterintuitive. "We have this crazy notion that river forecasting centers ought to be located on a river," a sarcastic Richard Hirn of the NWSEO quipped. In response, Hosler explained that river forecasting is done by interpreting data provided by river gauges, something that could be done anywhere. "There's no reason why a river forecasting center needs to be near a river. This could be done in Berlin," he said.[68] Nothing better demonstrates the naive faith that weather service boosters had in the far-reaching, almost magical ability of technology to obviate the need for direct human contact and observation of the natural world.[69]

CALL WAITING

One of Robert Eichhorn's worst nightmares almost came to pass in 1997. Proposed cutbacks at the NWS threatened to leave the National Hurricane Center (NHC) in Florida, responsible for tracking potentially deadly storms, with just one meteorologist on duty during the midnight shift, a worrisome prospect for Eichhorn, director of the New Orleans Office of Emergency Preparedness. Suppose a hurricane changed course in the middle of the night, threatening cities along the Gulf Coast, and emergency preparedness officials all started dialing into the hurricane center. "If they only have one person there, it's going to be backed up," Eichhorn said.[70] Let's hope somebody remembered to install call waiting.

The slated budget cuts were to reduce the number of positions at the NHC from 50 to 38. Five of the positions to be cut were meteorologists. Of course the lone person on duty during the midnight shift could always call in forecasters for overtime should an emergency arise. But we have already seen how well that strategy worked during the 1988 North Carolina tornadoes, when the person staffing the Wilmington weather office decided to tough out the situation. "The thing that concerns me is that we'll have a lot of tired people and we'll be stretching them thin," said Jerry Jarrell, the center's deputy director. "When you stretch people thin, they make mistakes. We'll get away with it for so long, but the big fear is that sooner or later, there will be something we'll miss."[71] Palm Beach County emergency manager B.T. Kennedy put it more bluntly: "They're going to be walking zombies, having been on duty for 12, 16 hours."[72]

Also scheduled to be cut in the same round of belt tightening was a data buoy stationed off the Florida coast in the Gulf of Mexico, an important source of information on wave heights used to predict coastal flooding. The buoy had been commissioned just four years earlier (1993) after a storm plowed into the Florida panhandle, killing 51.[73] Now the device was about to join the ranks of other inoperative buoys in U.S. waters, like the one off the coast of Seattle linked by some to the deaths in early 1997 of three coastguardsmen. The men had left on a rescue mission with a forecast of seas of 12 to 15 feet, but died when their boat capsized in waters roughly twice as high.[74] One anguished forecaster in Seattle pleaded with her superiors:

> I think forecasters working with inadequate data should be allowed
> to sign a waiver releasing them from any/all liability of forecasting
> under these conditions—that or be given the choice not to write a
> forecast due to inadequate data. Do the decision makers who have
> decided not to fix these buoys have any idea what a forecaster must
> feel when they blow a forecast so severely that a ship sinks and the
> entire crew is killed? Nobody in their right mind wants to live the
> rest of their life with that in the back of their minds.[75]

According to then NWS director Joe Friday, rough seas had prevented
the Coast Guard from servicing the inoperative buoy.[76] Still, Friday
empathized with the forecaster—noting her "serious emotional state of
distress over the deterioration of the NWS's infrastructure"—and sent
her message on up the ranks.[77]

"What we are seeing, in my professional judgment, is the equivalent
of losing the joints from your fingers," said Ronald McPherson, a top
weather service official who threatened to quit after almost four decades
with the NWS were the cuts to go through.[78] McPherson, along with
three other key people in the NWS management, wrote NOAA chief D.
James Baker that the proposed staff layoffs and budget cutbacks were
going to "jeopardize public safety by greatly increasing the risk of
weather-related disasters."[79] Robert Sheets, the former director of the
NHC, called the cuts "totally irresponsible."[80] And 26 past presidents of
the American Meteorological Society spoke out against them.

The proposed cuts at the hurricane center never came to pass, and
the Florida buoy at least was saved. Secretary of Commerce William
Daley decided to maintain the center's current level of funding but
promised, once the 1997 hurricane season had ended, to revisit the
staffing question at the NHC.[81] Meanwhile, the threat of the perpetually
hovering budget ax sent morale in the service to a record low. "I must tell
you that I have been in this organization now for 37 years, and I've never
seen morale at as low a point as it is now," said McPherson.[82]

Today's modernized National Weather Service has slimmed down to
121 forecast offices. All of those offices are to be stocked with one warning-
coordinating meteorologist doing community outreach, certainly in sum
an improvement over the handful of such personnel who once provided
the service. The offices have the latest in Doppler radar and are no doubt
better prepared technologically to face extreme natural forces then they

were in the past. But the service hangs under a Damocles sword, as each budget cycle rolls relentlessly onward and the threat of layoffs and cutbacks continues. Modernization has not changed any of this.

Those responsible for modernization have assumed that without new and better technology, Americans would not be able to counter extremes of nature. Nobody gave much thought to how chronic staffing shortages themselves were at the root of the natural disaster problem, and yet ample evidence exists that such human factors played a major role in contributing to the disasters of the 1970s and 1980s. One cannot help but conclude that in their zeal for the "remote control," those in charge of overseeing the nation's weather services have made a fetish of technology in their fight to predict nature. By subordinating the needs of a labor force to the imperatives of technology, allowing staffing matters to remain a chronic issue, and failing to fully address the human dimension of natural disaster, at times the government has helped to contribute to the problem it has set out to solve.

EIGHT

·······················

Who Pays?

There's a story about a fisherman from Pass Christian, Mississippi, who tried to save himself as wind and rain pitched him overboard during Hurricane Camille in 1969. About to be swept to his death, he somehow caught hold of a bridge railing and tied himself to it around the shoulders. There was only one problem: The rope was too short. When the high water receded, the rope, instead of depositing the fisherman to safety on the ground, crept slowly up his neck and strangled him, leaving him swinging in the summer wind.[1]

Which is worse, being swept to death by raging water or slowly hanged? The question must have occurred to more than a few of Mississippi's poor in the post-Camille months as federal and state officials stepped in to restore things to normal. The lesson of Hurricane Camille and many of the other disasters since the late 1960s (when the federal government first began playing a major role in providing relief to individuals) is this: Give government decision makers enough rope and they are destined to choke off any attempt to help the poor reach some semblance of economic justice in disaster's wake.[2]

Why is this? First of all, a faulty interpretation of calamity itself is in part to blame for the government's failure. Policymakers have tended to justify their actions in the realm of postdisaster relief aid by saying, in effect: Look at what nature has done to us.* Since natural forces are often unforeseen, political leaders have reasoned that it is impossible to budget adequately for disaster—a debatable point given the nation's past geophysical history. Woefully insufficient annual relief appropriations, in turn, have forced Congress to pass supplemental budgets, a highly charged political undertaking that has not always worked to the advantage of those most in need. Moreover, such chronic underfunding can be read as a collective act of denial that has, if nothing else, simply reinforced the unpredictableness of these events—creating, in turn, a huge self-fulfilling cycle of doom. Second, relief officials have tended to embrace a set of wrong-headed assumptions about the poor, about their style of life, their spending habits, their values, that, added to the usual bureaucratic incompetence, helps to explain why when the wind howls or the earth shakes, those least able to tolerate such shocks are pushed that much closer to the economic abyss.

USER-FRIENDLY AID

On Good Friday 1964, Alaska forwarded California something it could have lived without. Late in the afternoon, the Continental and Pacific plates ground past one another beneath the Gulf of Alaska, generating an M 9.2 earthquake, one of the most violent ever recorded in the history of the world. The quake caused Anchorage's main business section along Fourth Avenue to plunge 30 feet and set off a landslide that sent 70 homes in the upscale Turnagain Heights section skidding off their foundations. Meanwhile, 500 miles of ocean floor had moved violently upward in the quake, causing the water on top to mount into a tsunami. The wave, moving at a brisk pace of 400 to 500 mph, headed south, striking Crescent City, California, some 1,600 miles away. Eleven people died in the tidal wave.[3]

---

* Since the 1970s, relief aid has accounted for the largest share of federal disaster assistance, dwarfing mitigation funding by a factor of three.

Together, the earthquake and tsunami wound up dispatching 131 Americans, the start of what proved to be a deadly spate of natural disasters that over the next 18 months left almost no section of the country untouched. Hurricanes Hilda (1964) and Betsy (1965) pummeled the Southeast, combining to kill 113; a raging flood—nearly a 1,000-year event—traumatized the Northwest, leaving 45 dead; and tornadoes raked the nation's midsection, sending 256 to their graves. In all, no fewer than 545 Americans died as a result of natural disasters between March 1964 and September 1965, mocking the otherwise downward sloping curve of death that has come to define natural disasters in the twentieth century. And these were just the major calamities to strike the nation. In all, more than 200 congressional districts were ultimately declared disaster areas by the president.[4] It was not so much the death toll or the exorbitant property damage as the fact that so many Americans felt the effects of natural catastrophe during a relatively brief period that distinguishes this chapter in the annals of calamity.

This dramatic nationalization of disaster led some in Congress to wonder if the government was doing enough to help its citizens recover in the wake of such tragedies. Chief among them was Sen. Birch Bayh (D-Indiana). Bayh first became concerned with federal disaster relief after the 1965 Palm Sunday tornado outbreak fatally injured 138 of his constituents. He sponsored disaster legislation, passed in 1966, that made low-interest federal loans available to private homeowners and business people without the necessity of congressional approval. This hardly sounds like a significant legislative intervention, but it was. Understanding why requires a short detour into the history of federal disaster policy.

It is not often realized that no permanent means of government disaster assistance existed in this country until very recently. Only in 1950 did Congress pass legislation allowing the president to make disaster declarations to aid state and local governments in repairing public facilities (prior to this it required a special legislative enactment to receive such aid).[5] Even as late as 1969, no formalized means existed to help individual citizens in the wake of catastrophes. The reigning relief ethos was nicely summed up by George Hastings of the Office of Emergency Preparedness, then charged with executing all federal relief efforts, when he remarked in a moment of bravado, "We deal with things and not with people."[6] Otherwise, the needs of private individuals were to be tended to by the Red Cross, officially sanctioned by Congress in 1905 to deal with such matters. Not until Bayh sponsored the Disaster Relief Act of 1969 did such aid become more people friendly. The act allowed the federal bureaucracy, once the president had declared a disaster area, to extend all kinds of relief to John Q. Public. Now the Small Business Administration (SBA) could offer loans with a forgiveness clause that canceled the first $1,800, effectively making the loan into a partial grant-in-aid. The federal government offered temporary housing. It provided money for clearing debris from private property. It extended unemployment compensation. And it distributed food and food coupons to low-income people.[7] In short, Congress had woven for its citizens an entire disaster safety net.

In putting forth his disaster initiative, Bayh singled out angry Mother Nature as the problem that Congress had squared off to solve. "The sole purpose of this bill," he explained as he introduced the 1969 disaster act, "is to help alleviate the severe losses to property and livelihood so often inflicted on the unfortunate victims of unforeseen natural catastrophes." Again: "The purpose of this bill...is to help unfortunate victims, dealt cruel blows by entirely unexpected and unpredictable natural forces."[8] Seeing natural disasters as unforeseen acts of nature—as opposed to complex interactions between the natural world and social and economic forces—remains a common formulation that has repeatedly surfaced in congressional debates over disaster relief. Moreover, this interpretation has been used to rationalize the underfunding of the Presidential Disaster Relief Fund and the SBA since the 1970s.[9] After all, how can you budget for something you don't expect? The funding shortage forces Congress to pass a supplemental appropriation, a process fraught at times with partisanship and political wrangling. Meanwhile, disaster victims wait for help.

It's easier of course to generate bipartisan support for bankrolling a federal program when the evil in question—nature—is seemingly one faced by all Americans, rich and poor, black and white, Democrat and Republican. It's also easier to pass such legislation when nature is on the rampage. With Bayh's bill on the table, Hurricane Camille (category 5)—next to the Great Labor Day Hurricane of 1935, the most intense storm to ever strike the nation—roared through the South, killing 256 people and causing more than $5 billion in damage.[10] Camille landed its most powerful punch in lower Mississippi, a place buffeted by declining agri-

Biloxi, Mississippi, after Hurricane Camille's 20-foot storm surge (National Oceanic and Atmospheric Administration, Department of Commerce)

cultural prospects and mired in poverty, the effects of which were felt mainly by the region's black population. Spurred on by the disaster, Congress worked with tremendous speed to hammer out the details of the Disaster Relief Act, signed into law on October 1, 1969, precisely in time—had relief officials not been so scandalously disinclined to help the region's poor—to deal with the aftermath of the tragedy.

To say that the federal response to Camille in Mississippi was a disaster is to put it mildly. The list of abuses is lengthy. The Department of Housing and Urban Development (HUD) offered the homeless trailers, but agency rules required the recipient to provide a lot, something well beyond the reach of most poor black families made homeless in the calamity. HUD also did little to publicize the availability of temporary housing and gave people only 90 days of rent-free living when the law permitted as much as one year. Its reluctance to engage in relief was rivaled only by the SBA, which, in requiring substantial collateral for its loans—an unfair requirement since the forgiveness provision functioned effectively as a grant—discriminated against the poor. The SBA also failed to advertise its loan program, something private loan companies in the area—some charging interest rates as high as 40 percent—had managed to do. And, in an echo of Charleston 1886, its application forms were also needlessly complicated, a point of no small importance in the area affected by the disaster, where 39 percent of black males had just four years of education or less. "We don't have much patience with someone who can't fill out a form," remarked one SBA official.[11] That was simply another way of saying: poor blacks need not apply. And they didn't. Just 21 out of some 617 SBA loans went to nonwhites.[12]

Owen Brooks of the National Council of Churches, disturbed by the way disaster relief had gone about restoring Mississippi's poor to their former untouchable status, spoke the truth when he pointed out to Congress that life for such people, especially blacks, "is a year-round natural disaster." "It would be utopian I know to ask this subcommittee to recommend a law which would acknowledge this fact and redefine the term 'natural disaster' so as to provide poor people a decent job and income year around."[13] Perhaps nowhere was the artificiality of the natural disaster concept better demonstrated than in post-Camille Mississippi, making it all the more important for federal policymakers like Bayh to emphasize the unforeseen natural forces at work in such calamities. Otherwise, the entire concept might unravel before the appalling injustice of a social system in which human economic forces were capable of destruction as bad as, if not worse than, a category 5 storm.

Two years later, after the San Fernando earthquake (M 6.7) killed 58 and destroyed 800 structures, the federal disaster bureaucracy, the secret benefactor of middle-class America, again managed to dragoon nature into its campaign of keeping down the poor and elderly. The preferred method again turned out to be the forgiveness feature written into the loans. Although set up to function as a grant-in-aid, these were still technically loans, requiring recipients to be evaluated on their repayment ability. On such a technicality, many a rejection letter turned. Vannoy Thompson, a contractor from Los Angeles, told Congress after the 1971 quake that more than 90 percent of the SBA loan applicants from poor South Central were denied. Approval seemed to follow class lines. As he explained, "My doctors get their money. My businessmen, with a few exceptions, are getting their money. My lawyers get their money. But my John Does and my Jane Does didn't get their money, and that is what disturbs me."[14] Likewise, after Hurricane Agnes in 1972, the American

Friends Service Committee found that while the elderly "were not actually being denied loans, some were being given terms they could not afford; thus, in effect, they were denied access to the program."[15] Worse yet, in 1973, Congress, horrified by the cost of the relief bill and reports of the misuse of loan funds, eliminated the forgiveness provision (which had been raised to $5,000 in the wake of Agnes and the Rapid City flood that same year) and increased the interest rate from 1 to 5 percent. The changes further aided the well-off at the expense of the poor. Given Internal Revenue Service rules in the 1970s (allowing deductions for non-reimbursed losses and interest on disaster loans), the government effectively assumed a larger proportion of the losses suffered by those in the highest tax bracket.[16]

If the economics of natural disaster relief have tended to overlook those most in need, the point became abundantly clear in 1977, when spring floods devastated central Appalachia, a calamity brought on in part by reckless corporate strip mining. The coal companies bulldozed the landscape, sending tons of silt scattering into streams, raising their beds, and aggravating the effects of heavy rains. Over 22,000 people in four states—West Virginia, Kentucky, Virginia, and Ohio—experienced some kind of loss in the "natural" disaster. Again, the federal relief effort failed to come to the aid of those who needed help the most. "We've had two disasters here," remarked Jim Bartlett, a college professor from Mingo County, West Virginia. "One was the flood and the second was the federal relief effort."[17] HUD took forever to house people in mobile homes, in part because the agency initially chose to locate its assistance centers in Bluefield, West Virginia, roughly three to four hours driving time from the disaster area itself. Imagine spending hours on the road after all your worldly possessions have been swept away. Even worse, it took a long time to find HUD under *normal*—not flooded—driving conditions. Not to mention that many people found their cars destroyed in the calamity. If you managed to make it to an assistance center you would learn that the SBA no longer allowed forgiveness on its loans. And more bad news: The interest rate had ballooned by this time to more than 6 percent, almost the rate charged on conventional loans. All this for a house that perhaps did not even exist anymore! Although the SBA was authorized to grant up to three years before repayment began, in practice the agency offered victims just five months of grace. You can see why they call it the Small *Business* Administration.

Or the *Small* Business Administration, for that matter. Six weeks after the calamity, the SBA had interviewed over 4,700 possible applicants, but approved a minuscule 265 loans.[18] By all accounts, the agency's treatment of central Appalachia accorded well with the region's status in the 1970s as a national wasteland, a place pillaged for its resources but otherwise largely invisible to the collective social conscience. As one observer pointed out, "for Central Appalachians, calamity is a daily happening, not a one-shot occurrence."[19]

ED KOCH ASKS

These days, with academia as engaged with theory as it is, few would think to consult someone as pragmatic as Ed Koch for intellectual guidance. This is too bad, because the feisty former congressman and ex-mayor of New York City is capable at times of asking just the right questions. In 1972, while Congress was finishing up legislation to help the victims of the Rapid City flood and Hurricane Agnes, Koch (D-New York) rose to address his colleagues. If Congress was so willing to offer assistance to the victims of natural disaster, he mused, why was it so reluctant to offer similar relief to the sufferers of man-made calamities, people injured by an economic system that deprived them of decent housing and jobs. Koch continued:

> Would the ghettos of Harlem, South Bronx, Bedford Stuyvesant become eligible for the assistance now intended for the victims of Hurricane Agnes, if through some magical way we were able, without loss of life or personal injury, to have the East River flood those areas? Do we need the intervention of God before we address ourselves to the problems that man has created?

"I would like to know," Koch declared, "why it is we distinguish between natural disasters, and those made by man."[20] It was a good question.

To begin with, the federal system of disaster relief is set up to respond financially when a natural event causes physical destruction to property. In this sense, the response to disaster is biased in favor of the owners of wealth. Consider, for example, what happened in the spring of 1970, when South Florida was deluged with heavy rains, flooding farmland and putting an estimated 40,000 people out of work. Gov. Claude

Kirk succeeded in convincing Secretary of Agriculture Clifford Hardin to declare a natural disaster, providing financial assistance to farm owners and their families. But what about those who lost their jobs? Unemployment assistance, it turns out, is available only after the president declares a major disaster. In this case, Governor Kirk refused even to ask President Richard Nixon for such help, arguing that joblessness did not constitute a serious problem.[21] Let all workers beware: Joblessness is normal and expected under our economic system, and thus federal relief kicks in only when a sudden and unpredictable natural force causes enough destruction to demand the attention of the country's highest elected official.[22]

But what if the cataclysm in question is clearly man-made, not natural? Can the president declare for it? President Jimmy Carter decided to find out in 1978 when he classified the leaching of toxic chemicals at Love Canal a major disaster. Two years later he created another stir when he used the same tactic in dealing with the influx of Cuban refugees in Florida. Clearly these were not natural disasters in the conventional meaning of that phrase, and they broke with the custom in place for three decades. Between 1950, when Congress first authorized the president to declare major disasters, and the Cuban refugee crisis 30 years later, over 600 such declarations were made. Virtually all of them dealt with natural disasters, because that was how Congress worded the relief acts. The Disaster Relief Act of 1950 authorized the president to declare major disasters in the cases of flood, fire, hurricane, earthquake, drought, and storm. In 1970, Congress added high water, wind-driven water, and tidal wave to the list. In 1974, tsunami, volcanic eruption, landslide, mudslide, snowslide, and explosion were included. The phrase "other catastrophe" was also present in all the acts. Leaving aside the question of how much human responsibility to impute to these various calamities, it is clear that Congress chiefly intended its disaster taxonomy to deal with natural hazards.[23]

Of course the only natural thing about the Cuban crisis was the water in which the refugee boats floated as they made their way to Florida. Indeed, it's possible to see the entire sorry episode as not just man-made but government-made, when you consider the U.S. imposed trade embargo, the television and radio propaganda, and the fact that the Carter administration opened its arms to the fleeing Cubans not out of humanitarianism but to show the world that Castro lacked the support of his very own people. These were hardly unforeseen forces beyond human control. But that did not stop Carter from shamelessly stretching the meaning of "other catastrophe" to cover the Cuban refugees, a move that some fiscal conservatives in Congress denounced.[24] In response, they tried valiantly in the 1980s to police the boundary between nature and culture to foreclose on any further attempts to invoke disaster relief as a means of dealing with predictable social and economic problems.[25] Congress finally pushed through a more budget-friendly definition of calamity in 1988. Now a "major disaster" qualifying for a presidential declaration meant any "natural catastrophe (including any hurricane, tornado, storm, high water, wind-driven water, tidal wave, tsunami, earthquake, volcanic eruption, landslide, mudslide, snowstorm, or drought), or, regardless of cause, any fire, flood, or explosion."[26] Gone was the potentially treasury-draining phrase "other catastrophe." For the first time Congress explicitly stated that only natural catastrophes would qualify. The new definition did seem, however, to categorize some disasters as more natural than others. It singled out fire, flood, and explosion from the main list of perils and provided relief in these cases even where there was a strong man-made component to the destruction. This distinction further naturalized some calamities (hurricanes, tornadoes, earthquakes, and so on), though they too have important social and economic components at their heart. It also compromised the entire rationale for federal disaster relief, namely, that government intervention was necessary in the face of natural forces beyond human control.

The definitional life of calamity took yet another confusing turn in 1992. President George Bush declared disasters in Los Angeles and Chicago, but in neither case did nature play the chief role in the destruction. The Los Angeles riots were ostensibly motivated by the acquittal of white police officers accused in the Rodney King beating, though the stockpiling of class tensions provided much of the fuel for the explosion. (Another often overlooked motivating force was the execution of Robert Alton Harris the week before, as California resumed the death penalty.[27]) Thousands of buildings burned, creating a fire so vast and intense that it could be seen from outer space. In all, 60 people died. Yet some in Congress had trouble stomaching the idea of a presidential disaster declaration for an event so decidedly man-made, even if it technically qualified as a major disaster under the law. Congressman Norvell Emerson (R-Missouri), for one, objected to a federal bailout of the city. The Los

Angeles calamity, he said, "is a human disaster. Riots. Arson. Looting. Chaos. Killing. All of these were caused by human actors exercising their free will, not natural disasters."[28] Such a statement, however, glossed over the social and economic forces—high rates of unemployment, declining real wages, the resumption of capital punishment—that may also have contributed to the uprising, forces that were certainly beyond the rioters' control, even if they weren't "natural."[29]

The Los Angeles riots had the misfortune of coming on the heels of the Chicago flood, an event that many fiscal conservatives in Congress saw as simply an urban charity case, not a legitimate use of federal relief funds. The flood started when a minor leak in an abandoned tunnel beneath the city ruptured, sending more than 200 million gallons of water sprawling, causing a major blackout. Some in Congress felt the entire calamity could have been avoided. The city had known about the leaky tunnel for almost a year, said Frank Sensenbrenner (R-Wisconsin). Ten days before the deluge a city official, John LaPlante, received word that disaster would strike unless something was done quickly.[30] Following city procedure, however, LaPlante delayed repairs until he could solicit outside bids on the work. "I didn't get the feeling we had an immediate emergency," he explained later.[31]

It was this point about possible bureaucratic incompetence that some in Congress focused on when they labeled the disaster man-made. What they chose to overlook were the long-term human factors that contributed to the calamity. As anyone familiar with the city's history knew, Chicago had been built on a low-lying marsh, and flooding had long been a problem. The flood never would have happened if Chicago's founders had not chosen to build on the dividing line between the Mississippi and Great Lakes watershed. Nor would the flood have occurred had city authorities not elected to build the transportation tunnel in the first place.[32] None of these factors raised any eyebrows in Congress. And it seems safe to say that no one in the government would have objected to a federal bailout of flood-stricken Chicago had the allegations of bureaucratic ineptitude not come to light. In this sense, Congress employed a very narrow definition of moral responsibility for the calamity. It's clear that except for the bungling by city officials, Congress would have been content to funnel money to relieve the suffering caused by the flood, just as it had in other cases—earthquakes in California, hurricanes in Florida—where private-property-driven economic development, and not natural forces alone, contributed to the destruction. To call the Chicago flood a man-made disaster, as some in Congress did, was at one level a strategy for saving the government money. But such a categorization also tended to justify the expense of bailing out victims of what were seen as bona fide natural disasters, events like hurricanes and river floods, which also have—truth be told—an element of human complicity to them, but no obvious person or persons (as there were in the Los Angeles and Chicago cases) at whom to point the finger of blame. Finally, the 1992 disasters again demonstrated the queasiness of fiscal conservatives in Congress over providing relief in cases of seemingly man-made disaster. Such declarations, they argued, could send the government down the slippery slope of correcting the normal everyday injustices—racism, urban decay—of our social system.[33]

Thus we return to Koch's query: Why distinguish between natural disasters and man-made ones? Briefly put, it is a way for the federal government to subsidize prodevelopment interests and uphold the prevailing economic order, with all its injustice and convenience. The federal response to disaster is premised on bankrolling only those disasters the government deems sufficiently "natural" to warrant federal intervention. In practice, this has meant aid for victims of hurricanes, earthquakes, floods, tornadoes, and various other geophysical extremes. These are events beyond human control—unpredictable and unforeseen acts of nature or God—and it is only proper for the government to step in to right the situation, say liberals and conservatives alike.[34] Obscured, of course, are the human social and economic forces behind such calamities. Although the boundary between nature and culture is hardly a clear one, policymakers have behaved as though it were, doling out money to help relieve people when nature has rebelled, even as countless other economic disasters lacking a proximate natural cause go unaddressed. Of course, making this distinction helps to hide the fact that such "natural" calamities are as much if not more the product of a social order founded on the maximization of private property than they are the workings of the natural world. Meanwhile, certain man-made disasters are normalized and presented as simply ordinary events in no need of special funding or assistance. Such calamities, the argument goes, are just the price of doing business under capitalism. Let the line between the natural and the man-made fade, however, and you will find all too many unnatural disasters that merit fixing, all too many Harlems and Appalachias in need of help.

MEET FREDDY KRUEGER

In 1982, a student at Harvard asked Secretary of Defense Casper Weinberger whether he thought the world was going to end and, if so, how. "I have read the Book of Revelation," Weinberger responded, "and, yes, I believe the world is going to end—by an act of God, I hope—but every day I think that time is running out."[35] Soviet nuclear attack, not natural disaster, was the secretary's worst nightmare. He was far from alone in his concern. Precisely this vision of nuclear annihilation preoccupied his colleagues at the Federal Emergency Management Agency (FEMA). From 1982 forward, doomsters at the agency worried far more about mushroom, not funnel, clouds.

The government launched FEMA in 1979 after more than 30 years spent bouncing responsibility for disaster, natural and man-made, from one agency to another. In creating the agency, Carter had hoped to put to rest criticisms of the federal government's handling of emergency management. But under presidents Ronald Reagan and George Bush, FEMA was, according to a congressional report, quickly turned into "a dumping ground for political hacks."[36] Consider Phil May, appointed in 1981 to take charge of FEMA activities in the Southeast. Experience: six years working for a congressman and six months working to elect Reagan. Grant Peterson, appointed later to head the agency's disaster relief response, was a former television executive and county commissioner from Spokane.[37] Heading up the entire operation was Director Louis Giuffrida. A general in the California National Guard who had served as a longtime Reagan advisor, Giuffrida had written a bizarre thesis while at the U.S. Army War College in 1970. Titled "National Survival—Racial Imperative," the thesis dealt in part with the logistics of interning blacks in the event of a hypothetical urban riot.[38] Most of Giuffrida's background centered on responding to terrorist attacks and squelching political protests, which of course made him ideal to lead an agency ultimately concerned with responding to nuclear war. Bush himself appointed Wallace Stickney to head the agency, a man whose "only apparent qualification for the post," wrote Daniel Franklin in the *Washington Monthly*, "was that he was a close friend and former next-door neighbor of Bush Chief of Staff John Sununu."[39] In 1992, FEMA had nine times as many political appointees as the average for all federal government agencies.[40]

During the 1980s, FEMA spent the vast majority of its energies on an elaborate plan to safeguard the government during nuclear attack. The plan, established with the help of Col. Oliver North—famous for his grandiose schemes—eventually involved the deployment across the country of some 300 vehicles packed with generators, special radios, telephones, and satellite dishes. Rather than send the president and his aids to an underground bunker (the prior standard operating procedure during attack), the idea was to engage in a kind of shell game while still allowing government officials to communicate globally. Between 1982 and 1991, FEMA spent almost $3 billion developing equipment and plans for either protecting government officials during nuclear war or dealing with other aspects of national security. During the same time, it spent just $243 million on planning for natural disaster. In other words, FEMA spent 12 times as much on nuclear attack—ironically, the ultimate human-induced calamity—as it did on natural calamity.[41] "There is a holocaust mentality in the FEMA hierarchy which provides first priority to the nuclear attack-related activities and relegates all other natural and peacetime technological disasters to the back burner," explained Leon McGoogan, an Arkansas emergency management official in 1986.[42]

Still wrestling with the potential fallout of nuclear attack, FEMA was caught off-guard in 1989 when the United States began to experience a rise in mega-calamities. FEMA's expert preparedness for nuclear Armageddon, for example, hardly helped South Carolina's rural poor after Hurricane Hugo. Five days after the storm crashed through the state, federal relief officials had yet to provide food and water to victims in towns such as St. Stephen and Ridgeville, outside of Charleston. In rural Berkeley County, where St. Stephen is located, it took FEMA 10 days to open up a disaster assistance center. None of this is terribly shocking given that FEMA did not even establish a set of standard operating procedures to coordinate disaster relief efficiently. A full nine months after Hugo, some 1,200 families in rural South Carolina still needed help. Then, in 1991, FEMA abruptly shut down its funding of an outreach program designed to help victims both financially and emotionally. Outreach is critical to recovery in poor, rural areas where people are often unaware of the government programs set up to help them. The federal government just seemed to sit back and wait for people "to jump in their cars and ride down" into town for help, noted Mayor Robert Hoffman of St. Stephen. But as he pointed out, "a lot of people don't even know what FEMA is."[43]

While FEMA lurched into action in South Carolina, the Loma Prieta earthquake ripped through northern California in October of that same year. FEMA's response so appalled Alameda County Supervisor Don Perata that he wondered whether the IRS—routinely the butt of jokes about its mercenary nature—was getting a raw deal. "One used to think that the major horror in life in dealing with the Federal Government was the IRS; they have gotten a bad rap," he said. The IRS was "like Mary Poppins compared to FEMA as Freddie Krueger," the horror movie character.[44]

Again, FEMA failed to help those least able to withstand the trauma of a natural disaster. "The Bay Bridge gets rebuilt in a month, but the fate of these poor people gets lost and forgotten as the memory of the earthquake recedes," said Ken Zimmerman, a legal aid lawyer.[45] The poor people he had in mind were residents of the some 1,300 single-room occupancy hotels (SROs) destroyed in Alameda County alone. To the south, the city of Santa Cruz lost 500 rooms of SRO space, coming on the heels of over a decade's worth of steep cuts in federal support for affordable housing. The earthquake also devastated the largely Latino city of Watsonville, as well as Pajaro and Salinas, home to many of the region's migrant farmworkers. Predisaster life for these people was not exactly a picture of luxury. One grower in northern Monterey County charged workers rent to live in natural caves, fitted out with cardboard and tin partitions. A study of homelessness in the county showed that in some instances 40 to 50 people shared the same three-bedroom apartment, like a bad circus act, with workers, one after another, filing out of some Volkswagen-like flat and heading off to the fields.[46]

After the earthquake, FEMA entered ground zero to restore people to some semblance of their former lives. Given how far the poor had fallen, it must have seemed like a reasonable goal. But FEMA so botched the recovery effort that many indigent found themselves even worse off than before, as hard as it is to believe. Problem number one was FEMA's requirement that renters seeking temporary housing assistance produce evidence that they had lived in the same building for more than 30 days. Many residents of the region's SROs, however, were routinely forced out of their apartments every 28 days. Such temporary evictions prevented residents from attaining permanent status, making it easier for landlords to get rid of them in the future. Fewer than 100 people were denied housing

Downtown Watsonville, California, 1989 (Robert A. Eplett, California
Governor's Office of Emergency Services)

assistance under the 30-day rule, FEMA later claimed.[47] But this does not account for all those who decided not to seek aid because their understanding of the requirement made applying seem pointless.

Problem number two was FEMA's practice of discriminatory rental payments. Tenants in the region's residential hotels received one month's worth of assistance, while those who lived in anything but SROs were offered two months aid. Homeowners, moreover, were treated to a full three months of support. When Thomas Roberts of the Homeless Advocacy Project in San Francisco protested FEMA's treatment of the SRO population, he was told, "We know what they would do with the money if we gave it to them. They would smoke it or drink it."[48] In truth, there were lots of different kinds of people living in residential hotels, not just substance abusers. Some were there because they were underemployed. Some were retired. Some were physically disabled. And some simply could not afford the rent and security deposit required in the Bay Area's perennially tight housing market.[49]

When FEMA did turn down applicants for housing aid, they failed initially to inform people of their right to appeal the decision. If you have any questions call us, the letter said. In a masochistic moment, one legal aid worker actually did try to call FEMA, a task that required redialing the phone 114 times before connecting! One shudders to imagine calling FEMA from a phone booth out on the street. It is little wonder that more than half of the poor living in SROs did not receive housing assistance.[50]

In order to get FEMA to pay attention to the needs of low-income people in northern California, a coalition of 19 community groups began legal action against the agency. Early in 1990, FEMA agreed to settle the case by providing funds to replace "all SROs and shelter units, and buildings which were occupied and rendered uninhabitable by the earthquake."[51] The language here seems straightforward enough. But after legal aid attorneys submitted proof that more than 2,000 buildings needed such assistance, FEMA had a change of heart. It was all evidently a big misunderstanding. FEMA's attempt to back out of the agreement appeared so lame that the federal judge charged with disposing of the case had trouble concealing his amusement. U.S. District Judge Eugene Lynch wound up dismissing the agency's claims out of hand, calling them "suspect, to put it mildly."[52]

FEMA, it seems, made a habit of letting its bourgeois assumptions about the world obscure the very real needs of the poor, a point driven home again after Hurricane Andrew. "What we have learned is that FEMA has a pattern of coming into a disaster area, providing minimal relief and then folding up the tent," observed Terry Coble, an attorney with Legal Services of Greater Miami in 1992.[53] Two months after the hurricane disaster, FEMA had provided aid to fewer than half of the 154,000 people who had applied.[54] Even worse, it insisted on dispensing just one check per household, a policy based on middle-class norms inapplicable in places where extended family members or unrelated people teamed up to pay the rent.

That same year, victims of Hurricane Iniki faced similar inflexibility on FEMA's part, a remarkable failure if only because the ability to improvise is something that in an emergency should go without saying. The agency provided rent vouchers only to those dislocated Hawaiians directly affected by the disaster. Those who were evicted so that landlords could make needed repairs received no help until outrage over the policy forced FEMA to change it, some 10 months after the calamity. Noting FEMA's reluctance on this score, Kauai County Housing Agency director Chad Taniguchi pointed out that the island isn't the kind of place where "you can just get in a car and drive 50 miles if you need to."[55]

It is no accident that some of the worst federal relief operations have occurred among the poor. In part this is simply a question of political clout. But there are more fundamental reasons why the federal disaster response has failed the poor so miserably. It is simply very hard to figure out where the effects of a disaster begin and end. Obviously, in its attempt to back out of the low-income housing agreement, FEMA took a minimalist position. "We were set up to respond to emergencies not to build affordable housing," explained Tommie Hamner of FEMA's San Francisco office.[56] In other words, the agency was equipped to deal with extraordinary damage caused by unforeseen forces. Everyday man-made injustice was simply beyond its mandate. This sounds like a reasonable policy, and to some extent it is. But it is also a rationalization that implies, in this case, that the housing shortage was normal, the natural by-product of urban life.

PAY AS YOU FLOOD

After a decade of turbulence, FEMA began to settle down to more effective relief work in 1993, when James Lee Witt took over the agency. At least Witt, unlike his predecessors, had experience handling natural disasters. As director of Arkansas's Office of Emergency Services, he had acquitted himself well in the handling of three presidential disaster declarations, two that involved major flooding. Witt has refocused FEMA's mission, shifting it away from nuclear attack and toward natural disaster.[57] During the 1993 midwestern floods and the Northridge earthquake (M 6.7) the following year, the agency proved far more effective. Ironically, however, at just the point when the agency's response to natural calamity has improved, some in Congress have been dreaming up a more invidious way to ensure that the poor wind up paying for disaster—*literally* paying for it this time.

Congress's preferred means of loading the cost of disaster onto the backs of the poor has been the zero-sum budget politics that have come to define the 1990s. To understand how this new budget calculus affected the poor in times of need, it is necessary to take note of the Budget Enforcement Act of 1990. The act required Congress to find some source of revenue, either a budget cut or a tax hike, for all new federal outlays. There was, however, one exception. Emergency situations, such as natural disasters, were supposed to be exempt from the strictures of the pay-as-you-go approach. In other words, Congress was free to pass supplemental budgets funding disaster relief and simply add the cost to the federal deficit. This had long been the policy for dealing with natural disasters, and the new budget agreement did nothing to change it. But doubtless no one in Congress expected the spate of natural disasters that occurred beginning in 1989. Between 1980 and 1988, Congress had to pass only one supplemental budget for disaster aid in excess of $1 billion (though during the disastrous 1970s, *seven* such $1-billion packages were necessary). Who would have dreamed that in the period since 1989, Congress would see the need to pass no fewer than six such budget-busting appropriations to deal with the rising impact of natural disaster?[58] There were enough disasters to cause fiscal conservatives in Congress to worry whether the emergency provision of the 1990 budget agreement had been a big mistake.

The first sign that some in Congress had second thoughts about the budgetary status of natural disaster occurred after hurricanes Andrew and Iniki delivered their whopping financial blows. Ten senators voted against the more than $10-billion aid package.[59] In the House, Tim Penny (D-Minnesota) stood up and challenged his colleagues to think of how it would pay for relief, either through new taxes or offsetting budget cuts.[60] Few in Congress, however, seem to have paid him much attention.

Then the massive floods of 1993 struck. One might think that mid-western members of Congress would be the last to come forward and urge their colleagues to find a way to pay for relief aid when their constituents back home were battling to keep their heads above water. That said, meet Congressman Jim Nussle (R-Iowa). Seeking to fashion himself as a real fiscal tough guy willing to hold to the pay-as-you-go rationale no matter the depth of the water, Nussle urged Congress to find ways to offset the impending cost of natural disaster. And for one brief, extraordinary moment, he succeeded. The House actually temporarily blocked a $3-billion flood relief bill because it failed to include offsetting reductions, demonstrating an iron will to rein in the deficit.[61] But the victory was fleeting. With history on their side, Democrats—many from the Midwest—argued that never before had they been called on in emergency situations to identify ways of paying for relief aid. The relief bill passed, but not without the help of those in Congress eager to prove that the flooding was an act of God, an unforeseen, unpredictable event just like other such disasters the government had financed over the years.

After all, explained Sen. Robert Byrd (D-West Virginia), acts of God were exactly what Congress had in mind when it exempted emergencies from the budget agreement in the first place. Surely no one, he intoned, expected the victims of natural disasters to pay for these events themselves. Why? Because, simply put, "there are certain things that man cannot control." Hurricanes Andrew and Iniki, Typhoon Omar, the drought, freeze, and flooding of the early 1990s were all events that Congress paid for in one way or another, Byrd pointed out. And it paid for them because "every one of these emergency designations have been for an act of God. There was not a thing that any of the victims could have done to stop them."[62] Nor could anyone have done anything "to stop the rain in the Midwest over the last month or so." Maybe no one can stop it from raining, but surely the flooding at issue was not entirely God's doing, especially when you consider all the government-sponsored levee building and farming that destroyed vast tracts of wetlands, thus aggravating the destruction. Calling such events acts of God, as we have seen, has long been a way to evade moral responsibility for death and destruction. Only here such rhetoric was employed to assume financial responsibility, while shifting the actual cost to future generations of

American taxpayers. One of the most troubling things about those in Congress opposed to the assault on disaster relief is their eagerness to embrace a rhetorical stance destined to fail. In this respect at least, the fiscal conservatives are on stronger ground. As pointed out by Timothy Muris, a former budget official in the Bush administration, "rather than treat emergencies as acts of God," we should instead come to see that "these events are, in the aggregate, predictable."[63]

It was not, however, until 1994 that the pay-as-you-go contingent in Congress gained any real ground in their cause. The supplemental budget passed in the wake of the Northridge earthquake included more than $3-billion worth of recissions. These cuts were felt by departments across the board, from agriculture and defense to commerce, labor, and energy. Although the budget ax trimmed some money targeted for the poor—$10 million from a surplus food program, $3 million from legal aid—the cuts were not aimed at any one group in particular. But the relief bill was notable in at least one respect. The terms of the debate over how to pay for natural disasters had changed. A consensus had emerged that Congress refrain from adding disaster relief to the deficit. Now it was only a question of the size of the offsetting cut, which, in this case at least, turned out to be relatively modest.[64]

Still more ominous signs that fiscal conservatives in Congress were out to exact budgetary retribution on the poor emerged in 1995. That year, Congress passed the largest package of budget cuts in U.S. history as part of a $6-billion bill, the bulk of it for the Northridge calamity. This time the deficit hawks were able to exact cuts in social spending to pay for relief aid, cuts that fell disproportionately on the poor. Low-income housing, job training programs, and the Low Income Home Energy Assistance Program (LIHEAP), all felt the pain of the budget ax. HUD's program for housing the poor alone saw a reduction of more than $6 billion.[65] Renters in Los Angeles's poor sections and all across America were now effectively bearing the cost of rebuilding infrastructure and homes wrecked by the 1994 earthquake.

But nothing better demonstrates the cold-blooded class warfare at the core of the new pay-as-you-go paradigm than the cuts in the low-income energy program. The program benefits the poor, who spend roughly 20 percent of their income paying for utilities, a figure five times that of other households. President Bill Clinton sought to cut the program's budget, despite a chilling study done in 1992 showing that poor parents tended to withhold food from their kids to pay the heating bills.[66] Caught in a so-called heat or eat dilemma, bone-chilled residents of northern climes were brought to the brink of one kind of calamity in order to pay for another more visible one.

Meanwhile, with Congress putting the finishing touches on its cuts in low-income utility assistance, a deadly summer heat wave in Chicago showed, as never before, just how important such federal aid can be. It was as if some higher force was speaking up. Chicago lost at least 485 people (and maybe as many as 739) in just a week, more than the average number killed annually in the United States in all hurricanes, floods, and tornadoes combined. There were so many dead bodies that the Cook County Medical Examiner's office had to call in refrigerated tractor trailers, and even that move failed to solve the space shortage. Ambulances and police vans had to keep circling the morgue, desperately looking for a spot to unload yet another scorched victim.[67] Of course, it was not society's well-off—with their air-conditioned homes and cars—who wound up in a holding pattern at the morgue. It was the elderly, whose circulatory systems had slowed down; the poor, especially those living on the top floors of brick buildings with black tar roofs; and the estranged, who paid dearly.[68] A month after the hot spell ended, 41 bodies abandoned to the inferno remained unclaimed and were carted off to a mass grave, giving new meaning to urban anomie.

Although Chicago's medical examiner called the heat-wave deaths an "act of God," in truth (explains a NWS report), city officials, while well prepared for blizzards and floods, had ignored the threat of extremely hot weather.[69] As it turns out, however, Chicago had been burned before, most recently in August 1988, when a heat wave claimed an untold number of city residents. I say "untold" because the actual number of people who died in the disaster remains unclear, and for good reason. There is no federal definition of what constitutes a "heat death," and as a result the number of fatalities in this kind of event has been grossly underestimated.[70] One study conducted in 1972 observed that actual heat deaths are *10 times* the reported figure.[71] The lack of a uniform reporting system prompted W. Moulton Avery of the Center for Environmental Physiology in 1988 to point out that the government does a far better job tracking heat-related crop losses than human ones: "We are more concerned about corn than about our people."[72] No doubt this failure to properly reckon with heat deaths accounts in part for why Chicago in

1995 had a so-called heat plan totaling a monumental one-and-one-half pages.[73]

Which program did Clinton turn to for $100 million to soften the heat disaster's impact?[74] If nothing else, the entire dismal episode offered an object lesson on how important low-income energy funding can be. But it was a lesson lost on this country's highest leaders, who rushed into action—at exactly the time when people were frying to death—to cut LIHEAP, demonstrating yet again that blood and progress can flow in opposite directions.[75]

Perhaps even more worrisome than the actual funding cuts is the precedent that the 1995 budget process set. Prior to the passage of the budget, Congressman Newt Gingrich (R-Georgia) and the rest of the House leadership sent a letter to Clinton declaring that it would be their policy from now on to pay for natural disasters, presumably with budget cuts.[76] Instead of standing up to Republicans in Congress, Clinton caved in, conceding, as the *Congressional Quarterly* noted, "that the federal government should not go further into debt to assist victims of natural disasters."[77] And as long as the prevailing fear of raising taxes persists, the preferred method of payment for natural disasters may well be cuts in social programs.[78] A fundamental change in the political economy of risk has taken place. What the new budgetary calculus means is that the poor will pay twice for natural disaster, as they continue to be left out of the relief equation and then are made to bear the costs of that very same relief effort through cuts in social spending. As historian Mike Davis has observed, the entire scheme amounts to "an ingenious strategy for recycling natural disaster as class struggle."[79] Thus the forecast: high winds and rain with maybe a strong tremor giving way to increasing clouds— very dark ones—over Harlem, Appalachia, South Side Chicago, and the other unnatural disasters that litter the landscape of late-twentieth-century America.

EPILOGUE

....

Remembering
McKinneysburg

In March 1997, the small city of Falmouth, Kentucky, became yet another example of what happens when nature and governmental neglect intersect.[1] The city, with its 2,400 inhabitants, is located on the South Fork of the Licking River. Most of the time, the South Fork wends its way around the town in a wide arc, in no rush to meet up with the main stem of the river. There have been occasions, however, when the river has sought to shorten its trip, dispensing with the longer journey and instead forging its way directly through town. Nineteen sixty-four was one such occasion. So was 1997. It had been suggested as early as the 1930s that Falmouth ought to have a dam to protect it during such floods. But the dam was never built, mainly because the city, with its small population and limited economic potential, had routinely fallen prey to a cost-benefit calculus served up by Washington bureaucrats. Without a dam to protect them, Falmouth's residents had little choice but to rely on the National Weather Service's flood forecasts, alerting them as to when to evacuate. On Saturday morning, March 1, 1997, the weather service predicted a crest the next night at 12.5 feet above flood stage.* This was a prediction that many in the town, who had safely weathered the 1964 flood, figured they could handle. But the forecast was dead wrong. The river rose one foot per hour over the course of the day until chaos broke out near midnight as residents, lulled into a false sense of security, scrambled onto roofs and begged for help. The normally sedate South Fork was now raging at 100,000 cfs, cutting short the lives of five people, four of whom had checked into a shelter and then returned to their mobile homes to look after their property. (Emergency crews with German shepherds in tow later found the drowned victims floating in water that had filled their trailers to the ceiling.)

Although no hard evidence links the deaths with the botched forecast, at the very least, the last-minute evacuation could have been avoided had a stream gauge at McKinneysburg, Kentucky, roughly six miles upriver from Falmouth, remained open.[2] Used to monitor river levels, the gauge station operated for 56 years before being shut (along with 14 others in the state) in 1994 because of budget cuts.[3] When the decision was made to close the gauge, Monte Wheeler, a hydrologist with the weather service, wrote a letter of objection to the U.S. Army Corps of Engineers and the U.S. Geological Survey, which jointly operate the stations. "Without this hourly data," Wheeler explained, "the lead-time for warning the public of an impending flood in the city of Falmouth could be reduced by 24 hours."[4] Nobody paid Wheeler any attention, that is, until the 1997 flood struck. Of course there is no way to say for sure whether the McKinneysburg gauge, had it stayed open, would have saved lives. But clearly the gauge would have given the people of Falmouth a more accurate river forecast.[5]

McKinneysburg serves as a kind of negative symbol of how the federal government has failed its citizens in responding to natural disaster. Even more disturbing, what happened at McKinneysburg was not an isolated incident. Between 1983 and 1994, almost 400 gauge stations were shut down across the nation.[6] And this despite a 1983 government report indicating that the existing hydrological data network provided adequate coverage "for only one-half of the Nation's required river forecast

---

* In fact, it hit that mark early the following morning, and one day later had reached a record-breaking 25 feet.

Falmouth, Kentucky, 1997 (*Cincinnati Enquirer*)

points."[7] Recently, Gordon Eaton, the director of the Geological Survey, was asked how his agency goes about deciding which gauges to drop. Eaton likened the process to Russian roulette. The agency simply has to guess which ones are least likely to be needed. But as he quickly pointed out, "invariably sometimes you guess wrong."[8] It is hard to believe that the government would elevate roulette into public policy, but it is even harder to accept that chance alone governed the decision to abolish McKinneysburg, especially when the station helped to protect a poor city like Falmouth.[9] Surely, issues of class and power play a role in such decisions. In any case, there is no questioning the decline in gauge stations.[10]

The decision to mothball McKinneysburg put the weather service in the uncomfortable position of offering forecasts based on less than optimal information. But the closing of McKinneysburg and other long-standing gauge stations has had more far-reaching consequences. To begin with, such closings render the information painstakingly collected there for many years virtually meaningless, inhibiting the ability of forecasters to make predictions. Not only did people find their future foreclosed, but shutting gauge stations also helped sever people's relationship with the past. One might think of stream gauges as a kind of memory technology that offers precise records of flood heights. When such devices are destroyed, the attempt to remember the flood history of the river is reversed, setting the stage for forgetfulness—perhaps even for myth—to take hold. In this sense, the station closings are more than simply acts of budgetary brutality that stand in the way of more accurate forecasts. They are also acts of unconscionable historical violence.

When the lights went out at McKinneysburg, history and memory fell victim to indifference, just as did the people of Falmouth. Even though cutting the station resulted, ostensibly, from simple budgetary exigencies, the decision, whether it was meant this way or not, evinced a thoroughly ahistorical sensibility. It is now no longer possible to assay the river's present and future with respect to its past. Most distressing, by breaking the link between the present and the past, the fall of McKinneysburg is likely to obscure further the social and economic forces at work in flood disasters. Without an unbroken list of river readings, it will be even harder to discern how changes made in the name of economic development—deforestation, for example, or more paving done in concrete and asphalt—are likely to affect the watershed and future river levels.[11] The prospect of more land development can now take place unhindered by the troubling revelation of the role it has played in the creation of flood disasters. In the end, cutting such data-gathering points induces a kind of collective amnesia about calamity and plays into the worst impulses of denial.

The politics of forgetfulness, the subject of our tale here, will hardly aid the future response to natural disaster. To have seriously made the 1990s the Decade for Natural Disaster Reduction, as Congress mandated, would have required jettisoning such attempts to forget and deny, and embracing instead a new politics of remembrance. That new political framework should recall the past as a way of serving the present and future. It should strive to recapture the contingency of these events, instead of minimizing, denying, or naturalizing them. But to look squarely into the face of natural disaster, we must hold a mirror up to ourselves and the economic culture we have created. Only by coming to grips with the injustices and excesses of our social system can we hope to respond equitably and effectively to events that threaten the stability and integrity of that system.

Too often natural disaster has served as a kind of grand metaphor for rationalizing life under our economic system. These events just seem to happen. Comparing the 1994 Northridge earthquake with the earlier riots in Los Angeles, the political scientist Sherry Bebitch-Jeffe noted how the earthquake seemed "far less charged and value-laden."[12] But of course, as we have seen, there is a history and politics to "natural" disaster as vibrant and at times as vicious as any witnessed in riot-torn cities. Yet it is almost as if a giant magnet is rolled out after natural calamities, drawing people in power like so many iron filings into normalizing these phenomena and thus evading moral responsibility for their occurrence. This tendency to naturalize, to focus on nature and exaggerate its role in the destruction, amounts to one great exercise in moral hand washing. Even worse, by disavowing moral responsibility for disaster, we are rationalizing the kinds of economic oppression that explain why some people have stream gauges and others do not, why some people get adequate protection from floods and others cannot, why some live and some die.

The next time the Licking or some other river begins to roar, remember McKinneysburg, remember that there is no iron law of calamity, that disaster is not destiny, and, above all, that one person's act of God is— viewed from the perspective of history—just one more instance of man's inhumanity to man.

NOTES

....

ABBREVIATIONS

BAMS	*Bulletin of the American Meteorological Society*
BSSA	*Bulletin of the Seismological Society of America*
FEMA	Federal Emergency Management Agency
MDN	*Miami Daily News*
MH	*Miami Herald*
NC	*News and Courier* (Charleston, S.C.)
NOAA	National Oceanic and Atmospheric Administration
NWS	National Weather Service
NWSEO	National Weather Service Employees Organization
NYT	*New York Times*
OEP	U.S. Office of Emergency Preparedness
SCJ	*St. Charles (Mo.) Journal*
SFB	*San Francisco Bulletin*
SFC	*San Francisco Chronicle*
SFE	*San Francisco Examiner*
SLPD	*St. Louis Post-Dispatch*
WFO	Weather Forecast Office
WSO	Weather Service Office

INTRODUCTION: HOMETOWN BLUES

1. This statistic on flood heights in Hannibal was derived from table 4 on Mississippi River floods between 1851 and 1993 in J. Hurley and Roberta Hagood, *Hannibal Flood '93* (Hannibal, Mo.: n.d.), 153.

2. Construction company workers first tested the floodgates in March 1993.

3. Quoted in "Mississippi on the Mend," *U.S. News & World Report*, Aug. 22, 1994, 50.

4. The corps was pleased enough by its success to make the floodwall the frontispiece of its official report on the flood. See U.S. Army Corps of Engineers, *The Great Flood of 1993: Post-Flood Report*, Main Report (Chicago: Corps North Central Division, 1994).

5. The wall would have run roughly 6,200 feet and ended at the railroad yard south of the downtown. See *Report on Mississippi River Urban Areas from Hampton, Illinois, to Mile 300*, 87th Cong. 2d sess., 1962, H. Doc. 564, 85.

6. See U.S. Army Corps of Engineers, "Draft General Reevaluation Report for Flood Damage Reduction," Main Report with Draft Environmental Impact Statement (1985), 1:1.

7. Quoted in "Rising Spirits a Match for the Rising River Along Stretches of the Mississippi," *NYT*, July 4, 1993.

8. Quoted in "Flood of Indifference Overtakes Hannibal, MO., Family," *Columbus Dispatch* (Ohio), July 19, 1993.

9. Quoted in "People of the Mississippi," *Newsday* (New York), July 27, 1993.

10. On Twain the social critic, see Mark Twain, *A Pen Warmed-Up in Hell: Mark Twain in Protest*, ed. Frederick Anderson (New York: Harper and Row, 1979).

11. "Floodwall Proves Its Worth," *Hannibal (Mo.) Courier-Post*, June 29, 1993.

12. Professional historians themselves have not been terribly active students of natural calamity. See, e.g., John C. Burnham, "A Neglected Field: The History of Natural Disasters," *Perspectives* (American Historical Association newsletter), Apr. 1988, 22–24. Such neglect may have existed for some time. As far back as 1928, C.F. Talman wrote, "It is a paradox that contemporary historians, who pride themselves upon having broken away from the old conception of history as a record of wars and politics, almost completely neglect the history of natural disasters." See C.F. Talman, "Storms Have a Hand in History Making," *NYT Magazine*, July 29, 1928, 16. There have been, of course, some exceptions to this general tendency to ignore calamity. But most tellingly, it has been those disasters with obvious human components—floods and dust storms, for example—that have garnered the most attention from historians. See, e.g., Pete Daniel, *Deep'n As It Come: The 1927 Mississippi River Flood* (New York: Oxford Univ. Press, 1977); David McCullough, *The Johnstown Flood: The Incredible Story Behind One of the Most Devastating "Natural" Disasters America Has Ever Known* (New York: Simon and Schuster, 1968); and Donald Worster, *Dust Bowl: The Southern Plains in the 1930s* (New York: Oxford Univ. Press, 1979). For a stimulating discussion of the role of natural disaster in European and Asian history, see E.L. Jones, *The European Miracle: Environments, Economics and Geopolitics in the History of Europe and Asia* (Cambridge: Cambridge Univ. Press, 1981), 22–41.

The broader field of disaster research emerged in the 1920s. Perhaps the most important early work was Samuel Henry Prince's *Catastrophe and Social Change* (New York: Columbia Univ. Press, 1920), a sociological study of the collision between two ships—one carrying munitions—in Halifax Harbor in 1917. However, it was not until after the Second World War that the field of disaster research, as it came to be called, developed in earnest, responding to fears of the effects of an attack on the United States. Again, sociologists played a major role in leading the way. See E.L. Quarantelli and Russell R. Dynes, "Response to Social Crisis and Disaster," *Annual Review of Sociology* 3 (1977): 23–49; and G.A. Kreps, "Sociological Inquiry and Disaster Research," *Annual Review of Sociology* 10 (1984): 309–330.

While most sociologists have concerned themselves with the behavior of individuals and groups under the stress of disaster, geographers, who have also played a leading role in the postwar disaster-research initiative, have tended to focus on the geophysical forces responsible for natural calamities. In the 1970s, a spate of disasters (floods in eastern Pakistan in 1970, Hurricane Fifi in Honduras in 1974, the Guatemalan earthquake of 1975, and the T'angshan earthquake in China the following year) helped to highlight the problems of natural calamities and drew scholars, especially social geographers and anthropologists, to reconsider the prevailing assumptions of the dominant disaster paradigm, which had been forged, as indicated above, in the postwar years. See, e.g., Kenneth Hewitt, ed., *Interpretations of Calamity: From the Viewpoint of Human Ecology* (Boston: Allen and Unwin, 1983); Kenneth Hewitt, *Regions of Risk: A Geographical Introduction to Disaster* (Essex, U.K.: Longman, 1997); and Anthony Oliver-Smith, "Disaster Context and Causation: An Overview of Changing Perspectives in Disaster Research," in *Natural Disasters and Cultural Responses*, ed. Anthony Oliver-Smith, publication 36 of *Studies in Third World Societies* (1986): 1–34.

13. Kenneth Hewitt, "The Idea of Calamity in a Technocratic Age," in *Interpretations of Calamity*, 5, 10, 16. The op-ed page of the *New York Times*, for example, is often given over to expressions of the dominant "natural" view of calamity. After Hurricane Andrew, one writer ridiculed those who blamed the disaster on anything but natural forces. "Only Americans seem to believe that they have constitutional protection against the forces of nature." Similarly, after a very powerful tornado leveled a subdivision in Jarrell, Texas, killing 27, another writer offered that, unlike other natural disasters, "the confrontation [with tornadoes] is more haphazard and personal." While admitting that there are certain steps that the public could take to avoid future tornado fatalities (e.g., making people more aware of the difference between a "watch" and a "warning"), ultimately, "all you can do is cross your fingers." In fact, as will be explained further below, there was a lot more the NWS and the federal government more generally could have done to reduce tornado-related deaths. See "Hunting for Scapegoats in South Florida," *NYT*, Sept. 2, 1992; and "Twist of Fate," *NYT*, May 30, 1997.

14. J.M. Albala-Bertrand, *Political Economy of Large Natural Disasters: With Special Reference to Developing Countries* (Oxford: Clarendon Press, 1993), 204.

15. "1 in 5 See Floods as God's Wrath," *SLPD*, July 23, 1993.

16. Thomas Foxcroft, *The Voice of the Lord, From the Deep Places of the Earth* (Boston: S. Gerrish, 1727), 25.

17. David D. Hall, *Worlds of Wonder, Days of Judgment: Popular Religious Belief in Early New England* (New York: Knopf, 1989), 71–72, 78, 94.

18. Maxine Van de Wetering, "Moralizing in Puritan Natural Science: Mysteriousness in Earthquake Sermons," *Journal of the History of Ideas* 43 (1982): 422, 436. For more on earthquakes in the colonial period, see Michael Nathaniel Shute, "Earthquakes and Early American Imagination: 'Decline and Renewal in Eighteenth-Century Puritan Culture'" (Ph.D. diss., Univ. of California, Berkeley, 1977); William D. Andrews, "The Literature of the 1727 New England Earthquake," *Early American Literature* 7 (1973): 281–294; and Charles Edwin Clark, "Science, Reason, and an Angry God: The Literature of an Earthquake," *New England Quarterly* 38 (1965): 340–362. For reflections on the religious approach to disaster more generally, see Russell R. Dynes and Daniel Yutzy, "The Religious Interpretation of Disaster," *Topic* 5 (1965): 34–48.

19. James Lal Penick Jr., *The New Madrid Earthquakes*, rev. ed. (Columbia: Univ. of Missouri Press, 1981), 117–118.

20. My thinking here was influenced by Ulrich Beck, *Risk Society: Towards a New Modernity*, trans. Mark Ritter (London: Sage Publications, 1992).

21. "Winds of Chaos," *Time*, Oct. 2, 1989, 16.

22. Sparks discusses the wind-speed issue in "Lessons of Savage Storm Are Written on the Wind," *Atlanta Journal and Constitution*, Sept. 3, 1990. See also table 1 in Peter R. Sparks, "The Facts About Hurricane Hugo—What It Was, What It Wasn't and Why It Caused So Much Damage," in *Hurricane Hugo One Year Later*, ed. Benjamin L. Sill and Peter R. Sparks (New York: American Society of Civil Engineers, 1991), 279. The estimated fastest-mile wind speed for the Charleston area (none of the anemometers survived in those areas with the greatest winds) was 95 mph, with an approximate recurrence interval of 50 years.

23. See, e.g., "Building Codes Remain in Flux," *Post and Courier* (Charleston, S.C.), Sept. 21, 1994.

24. Peter R. Sparks, "Development of the South Carolina Coast 1959–1989: Prelude to a Disaster," in *Hurricane Hugo*, ed. Sill and Sparks, 3–7.

ONE · LAST CALL FOR JUDGMENT DAY

1. Quoted in Oliver Pilat and Jo Ranson, *Sodom by the Sea: An Affectionate History of Coney Island* (Garden City, N.Y.: Doubleday, 1941), 149.

2. The above quotations are from Richard Snow, *Coney Island: A Postcard Journey to the City of Fire* (New York: Brightwaters Press, 1984), 46.

3. John F. Kasson, *Amusing the Million: Coney Island at the Turn of the Century* (New York: Hill and Wang, 1978), 72.

4. By "business class," I mean those leading members of the commercial community, as well as their associates in the worlds of law and journalism, who supported

them in boosting a city's economic prospects. Although this concept has been used by many scholars over the years, the best, most concise definition is found in Don H. Doyle, *New Men, New Cities, New South: Atlanta, Nashville, Charleston, Mobile, 1860–1910* (Chapel Hill: Univ. of North Carolina Press, 1990), xiv. I will use "business class" interchangeably with "urban elite" and "urban leaders."

5. The following disaster chronology is based on Walter J. Fraser Jr., *Charleston! Charleston!: The History of the Southern City* (Columbia: Univ. of South Carolina Press, 1989), 11, 16–17, 43, 68, 84, 189, 216, 233–234, 253–254, 315; and John C. Purvis, *South Carolina Hurricanes, Or a Descriptive Listing of Tropical Cyclones That Have Affected South Carolina* (n.p.: South Carolina Civil Defense Agency, 1964), 9, 11, 12–13, 14, 17.

6. Chimneys were damaged in Savannah, 90 miles from Charleston, and a dam fell victim to the shock in Augusta, Georgia. See Robert P. Stockton, *The Great Shock: The Effects of the 1886 Earthquake on the Built Environment of Charleston, South Carolina* (Easley, S.C.: Southern Historical Press, 1986), 17, 30; and Jerry L. Coffman, Carl A. von Hake, and Carl W. Stover, eds., *Earthquake History of the United States* (Boulder, Colo.: Department of Commerce, NOAA, and Department of Interior, U.S. Geological Survey, 1982), 25, 27, 28.

7. Stockton, *Great Shock*, 29–30. Stockton says that in all, 76 people died in the calamity. But Karl Steinbrugge, a seismic hazards expert, asserts that the figure was 110. See table 1.2 in Karl V. Steinbrugge, *Earthquakes, Volcanoes, and Tsunamis: An Anatomy of Hazards* (New York: Skandia America Group, 1982), 5.

8. Fraser, *Charleston!*, 316; Stockton, *Great Shock*, 36–37.

9. "The City Full of Visitors," *NC*, Oct. 4, 1886.

10. "A Quiet Night," *NC*, Sept. 4, 1886. On September 6, 1886, the Knights of Labor held a meeting in Charleston and agreed to increase the wages of manual laborers by 50 cents per day. Bricklayers thus received $3.50 per day. Two weeks later, bricklayers commanded between $5 and $8, and that wage, for some, reportedly increased to $10 within two months of the calamity. See Stockton, *Great Shock*, 88.

11. Doyle, *New Men*, 9, 53–54.

12. Quoted in "Courtenay in New York," *NC*, Sept. 7, 1886. On Sept. 8, 1886, the *News and Courier* claimed that the interview it had published the day before, which was excerpted from the *New York Tribune*, was a fabrication, though there is no evidence for this, nor any reason to doubt the authenticity of the interview. See "'Courtenay in New York,'" *NC*, Sept. 8, 1886.

13. "A Quiet Night."

14. Ibid.

15. This quotation and the one directly above it are from "The Terror of the Night," *NC*, Sept. 5, 1886.

16. Doyle, *New Men*, 292–93.

17. Bernard E. Powers Jr., *Black Charlestonians: A Social History, 1822–1885* (Fayetteville: Univ. of Arkansas Press, 1994), 101, quotation from p. 227.

18. Kenneth E. Peters and Robert B. Herrmann, eds., *First-Hand Observations of the Charleston Earthquake of August 31, 1886, and Other Earthquake Materials: Reports*

of W.J. McGee, Earle Sloan, Gabriel E. Manigault, Simon Newcomb, and Others
(Columbia: South Carolina Geological Survey, 1986), 91.

19. "Charleston's Calamity," *Baltimore Sun*, Sept. 6, 1886. See also "Another Shock,"
ibid., Sept. 9, 1886.

20. "The Preachers and the Scientists," *NYT*, Sept. 8, 1886.

21. One newspaper account noted that while Charleston's white community was very
"impressed" with the seriousness of the calamity, the city's black population was
"completely unnerved and unstrung." In the aftermath of the quake, the black
population fled into the streets and "filled the air with dismal groans of despair
and lamentations." Running in order to avoid the "blinding clouds of pulverized
mortar," they prostrated themselves before God, yelling "Do my Master Jesus
have mercy on me" and "Let me live through this night, dear God, my Saviour."
The earthquake evoked an outbreak of "superstitious fear" among Charleston's
blacks that was "impossible to exaggerate." On Marion Square, six "colored
boys...had fallen to the ground in a paroxysm of religious frenzy. They were
grovelling with their faces down in the grass, and were singing a hymn in a loud
voice. The hymn was 'The angels a-rappin' at the door.'" See "A Day of Gloom,"
NC, Sept. 3, 1886; and "Scenes on the Streets," *NC*, Sept. 4, 1886.

22. "The Earthquake in St. Andrew's," *NC*, Sept. 3, 1886.

23. "All over the State," *NC*, Sept. 5, 1886.

24. "Elsewhere in the State," *NC*, Sept. 3, 1886.

25. Perhaps the authorities were already impressed enough with how the disaster had,
in the words of the black ministers, "come to cause every one to consider his way
and turn to the Lord." *Journal of the City Council of Charleston, S.C.*, Regular
Meeting, Oct. 12, 1886, p. 335 (microfilm, Charleston County Library).

26. "The Earthquake," *Christian Recorder*, Sept. 30, 1886. I was unsuccessful in locat-
ing a single black newspaper published in Charleston or nearby at the time of the
earthquake.

27. Peters and Herrmann, *First-Hand Observations*, 113.

28. Powers, *Black Charlestonians*, 216–17. On slave religion during the antebellum
period in the Low Country more generally, see Charles Joyner, *Down by the
Riverside: A South Carolina Slave Community* (Urbana: Univ. of Illinois Press,
1984), chap. 5, especially pp. 142, 160–61, 169; and Margaret Washington Creel, *"A
Peculiar People": Slave Religion and Community-Culture Among the Gullahs* (New
York: New York Univ. Press, 1988).

29. "A Word in Season," *NC*, Sept. 10, 1886.

30. In the postbellum period, whites sought to impose a system of wage labor on
blacks throughout the Low Country. That strategy brought them into direct confrontation with blacks, especially those who worked on the region's coastal plantations, who were used to a task-oriented approach to labor. This task orientation,
as historian John Scott Strickland notes, gave blacks a measure of independence
and control over the work process that made it all the harder for them to accept
the strictures of labor discipline that whites sought to impose, earthquake or not.
See John Scott Strickland, "Traditional Culture and Moral Economy: Social and

Economic Change in the South Carolina Low Country, 1865–1910," in *The Countryside in the Age of Capitalist Transformation: Essays in the Social History of Rural America*, ed. Steven Hahn and Jonathan Prude (Chapel Hill: Univ. of North Carolina Press, 1985), 144–147.

31. "The Gospel of the Quake," *NC*, Sept. 13, 1886. According to a summary of Porter's remarks given on the second Sunday after the earthquake, he explained that "by the force of circumstances we are peculiarly located in the course of nature for good and evil things, and that is all there is in it."

32. Quoted in "Another Slight Tremor," *Baltimore Sun*, Sept. 9, 1886.

33. "How to Live Now!" *NC*, Sept. 10, 1886.

34. "The Man and the Hour," *NC*, Sept. 8, 1886.

35. "How to Live Now!" Clearly, the city's business class construed the outdoor, carnivalesque atmosphere as a profound threat to normal life. However, city officials refused to give the appearance that order had broken down completely, lest the city's commercial prospects be compromised. At one point, having heard about the general "demoralized" condition of the lower classes in Charleston, the Brooklyn post of the Grand Army of the Republic volunteered its services to restore order. But the city rejected the offer and tried to downplay the trouble. "There has not been, at any time," read one report, "the slightest foundation for the rumors that have been circulated as to a condition of popular disorder prevailing in the city." See "Not Demoralized," *NC*, Sept. 11, 1886.

36. On mobility and black working-class protest in the Jim Crow South, see Robin D.G. Kelley, *Race Rebels: Culture, Politics, and the Black Working Class* (New York: Free Press, 1994), 25.

37. Quoted in E. Culpepper Clark, *Francis Warrington Dawson and the Politics of Restoration: South Carolina, 1874–1889* (University, Ala.: Univ. of Alabama Press, 1980), 138.

38. Quoted in ibid., 139.

39. For a discussion of Dawson's views on economic progress, see Doyle, *New Men*, 58, 111, 306.

40. "A 'Visitation,'" *NC*, Sept. 19, 1886.

41. Fraser, *Charleston!*, 203.

42. Wilbert L. Jenkins, *Seizing the New Day: African Americans in Post–Civil War Charleston* (Bloomington: Indiana Univ. Press, 1998), 51.

43. "Vox Populi, Vox Diaboli," *NC*, Sept. 7, 1885. Such a tough-minded approach to disaster was not unique to Charleston. Officials in other cities minimized disaster and turned down offers of outside help, even as late as the 1920s. For example, after a devastating tornado disaster in St. Louis in 1896, city officials refused to solicit outside funds even though the calamity caused massive destruction and unemployment. According to the St. Louis health commissioner, the city needed to demonstrate its self-reliance in the face of disaster. "As to asking aid outside of St. Louis," he said after the calamity, "that is all sheer nonsense. St. Louis can take care of its own people and does not want help. We want to be perfectly independent and self-reliant in this matter." Even several days after the disaster, as reports

of severe destruction emerged, Mayor Cyrus Walbridge insisted that St. Louis could handle the disaster on its own. Similarly, following a 1906 storm that severely damaged Pensacola, Florida, the mayor refused outside relief from Montgomery and other cities, preferring to downplay the suffering, which was in fact significant. Quotation is from "A Self-Reliant People," *SLPD*, May 31, 1896. See also "Mayor Hung in Effigy," *SLPD*, June 3, 1896; and "Rebuilding of Pensacola Continues with a Rush," *Daily Picayune* (New Orleans), Oct. 5, 1906.

Concerned that Charleston's leaders might not be inclined to solicit outside aid, a group of black ministers (including all of those who petitioned the city council for a fast day) felt compelled after the 1886 quake to make an appeal for aid on their own. They established a committee to receive and distribute relief funds to the city's blacks, a move criticized by Dawson's *News and Courier*: "Rich and poor, ignorant and learned, white and black are on the same footing in face of the common misfortune, and a heavy responsibility will rest upon persons who seek in any way to separate or drive apart those whom a terrible misfortune has brought close together." The ministers, wrote the paper, might better spend their time by seeing "that the charity of the American people shall not be imposed on, and that persons who are able to work, but who are unwilling to support themselves, and prefer to live in idleness, shall not be recommended as proper recipients of public aid." See "The Colored Clergy," *NC*, Sept. 7, 1886; and "No Sectional Line—No Color Line," *NC*, Sept. 8, 1886.

44. "The Work of Relief," *NC*, Sept. 8, 1886.
45. "Feeding the People," *NC*, Sept. 7, 1886. Attempts to starve blacks into working date from as early as 1865, when the South Carolina Freedman's Bureau issued a directive making clear that rations were only to be dispensed in extreme situations, and certainly not when people were capable of working. John Scott Strickland notes that this policy was used to discourage independent black agriculture in the Low Country. See Strickland, "Traditional Culture and Moral Economy," 153.
46. William A. Courtenay to William E. Dodge, Oct. 13, 1886, in William Ashmead Courtenay, *Charleston Cyclone and Earthquake Scrapbooks* (Charleston: South Carolina Historical Society, 1885–1887). The calamity seemed to offer many opportunities for those on the bottom of the social hierarchy to take advantage. A month after the disaster, it was reported that "idle colored people" continued to descend on the city in order to receive free rations. "These men have thronged the streets from day to day and have, in many instances, refused to accept all offers of employment." There were calls for the rigid enforcement of the vagrancy law. See "Work, or go to Jail," *NC*, Sept. 29, 1886.
47. *Journal of the City Council of Charleston, S.C.*, Special Meeting, Oct. 5, 1886, p. 333.
48. "Charleston Letter," *Christian Recorder*, Nov. 11, 1886.
49. A copy of the relief application can be found in Courtenay, *Scrapbooks*.
50. "Charity's Hard Questions," Letter to the Editor, *New York Herald*, Sept. 20, 1886.
51. Department of Interior, Census Office, *Compendium of the Tenth Census*, pt. 2 (Washington, D.C.: GPO, 1883), 1653.

52. "The Work of the Relief Committee," *NC*, Sept. 22, 1886.
53. Stockton, *Great Shock*, 72.
54. "The Shake Among the Clocks," *NC*, Sept. 11, 1886.
55. "Another Shock," *Baltimore Sun*, Sept. 4, 1886.
56. "A Day of Gloom." Similarly, when a tornado struck St. Louis in 1896, killing 225 people, reporters remarked on how the storm stopped the city's public clocks. See "What a Cyclone's Attack on a Great Community Means," *Harper's Weekly*, June 13, 1896, 594.
57. Michael O'Malley, *Keeping Watch: A History of American Time* (New York: Viking, 1990), 149.
58. This effort to normalize everyday life in late-nineteenth-century America also extended into the realm of money. In the period after the Civil War, the federal government sought to standardize money, culminating in the Gold Standard Act of 1900. See Viviana A. Zelizer, *The Social Meaning of Money* (New York: Basic Books, 1994), 14.
59. "An Unsafe Planet," *Daily Morning Chronicle* (San Francisco), Oct. 23, 1868.
60. "The Earthquake Year—Our Installment," ibid., Oct. 22, 1868.
61. This effort to normalize disaster was threatened, at least in part, by a parallel development in the history of calamity: the rise of natural disaster as a form of mass entertainment. At the same time that newspapers and the elite interests they generally represented were draining calamity of any deeper religious meaning, ordinary Americans were delighting in the horrors of life turned upside down. Tornadoes attracted a great deal of attention because their power could create seemingly bizarre results. The 1896 St. Louis tornado caused massive destruction, perhaps explaining why it surpassed all previous tornado disasters for its crowd appeal. On just a single day, 50,000 curiosity seekers descended on the city from as far as 250 miles away. The railroads, which had lost money during the storm, tried to recoup their losses by selling so-called tornado tickets, low round-trip excursion fares. In the three days after the disaster, the trains into the city were jammed to capacity. Hotels filled up and hostelries had to resort to cots to accommodate all the guests. Of course, some came to look after relatives and friends. But according to the *St. Louis Post-Dispatch*, "most of the strangers were drawn here by morbid curiosity." The devastated southern part of the city, especially Lafayette Park, with its huge uprooted trees and twisted iron fences, emerged as a primary tourist attraction. A carnival atmosphere, replete with vendors selling beer and snacks, and exhibitions (including one involving "the Roentgen X-rays"), sprung up to entertain the visitors. One week after the storm it was estimated that 150,000 people had passed through the city. The huge number of people created crowd control problems, especially at the city's main train station, and also interfered with attempts to rescue people and repair the damage. "St. Louis Burying Its Dead," *New-York Tribune*, June 1, 1896; "Lafayette Park," *SLPD*, June 6, 1896; "Sunday Visitors," *St. Louis Republic*, June 8, 1896. Quotation is from "Throngs of Visitors," *SLPD*, June 3, 1896.
62. "The New Pompeii," *New York Daily Tribune*, May 11, 1902.
63. "Earthquakes Then and Now," *Nation*, Feb. 4, 1909, 105.

64. Ibid. Emphasis added.

65. J.M. Nau and A.K. Gupta, "The Earthquake Threat and Its Mitigation in the Southeastern United States," in *A Workshop on "The 1886 Charleston, South Carolina, Earthquake and Its Implications for Today,"* ed. Walter W. Hays and Paula L. Gori, comp. Carla Kitzmiller (Reston, Va.: U.S. Geological Survey, 1983), 239, 243.

66. Majorie R. Greene and Paula L. Gori, "Earthquake Hazards Information Dissemination: A Study of Charleston, South Carolina," Open File Report 82-233, (Reston, Va.: U.S. Geological Survey, 1982), 7, 26.

67. *Earthquake Hazards, Risk, and Mitigation in South Carolina and the Southeastern United States* (Charleston: South Carolina Seismic Safety Consortium, 1986), 100.

68. Quotation is from Lyons's testimony before Congress. See House Committee on Science and Technology, *Earthquakes in the Eastern United States*, 98th Cong., 2d sess., 1984, Committee Print, 98–99.

69. Earthquakes are about 10 times more common in California than in the East. See "Earthquakes in the Eastern United States," *Earthquake Information Bulletin*, Nov./Dec. 1984, 212.

70. *Earthquakes in the Eastern United States*, 134, 164.

71. Quoted in Fred Graver and Charlie Rubin, "Waiting for the Big One," *New York* magazine, Dec. 11, 1995, 45.

72. Carl M. Wentworth, "The Changing Tectonic Basis for Regulatory Treatment of the 1886 Charleston, South Carolina, Earthquake in the Design of Power Reactors," in *A Workshop on "The 1886 Charleston, South Carolina, Earthquake,"* 268.

73. *Earthquakes in the Eastern United States*, 204–205; Wentworth, "Changing Tectonic Basis," 269.

74. "Had the 1886 Charleston earthquake not occurred during historic time," geologists Mark and Mary Lou Zoback have observed, "there would really be no reason to suspect it as a site of a potential major earthquake." See Mark D. Zoback and Mary Lou Zoback, "*In Situ* Stress, Crustal Strain, and Seismic Hazard Assessment in Eastern North America," in *Earthquake Hazards and the Design of Constructed Facilities in the Eastern United States*, ed. Klaus H. Jacob and Carl J. Turkstra (New York: New York Academy of Sciences, 1989), 63.

75. *Earthquakes in the Eastern United States*, 204–205.

TWO · DISASTER AS ARCHETYPE

1. Kenneth Hewitt, "The Idea of Calamity in a Technocratic Age," in *Interpretations of Calamity: From the Viewpoint of Human Ecology*, ed. Kenneth Hewitt (Boston: Allen and Unwin, 1983), 11–12.

2. Mircea Eliade, *The Myth of the Eternal Return or, Cosmos and History*, trans. Willard R. Trask (Princeton, N.J.: Princeton Univ. Press, 1954), 46.

3. Two on-line searches of the 38 million records available in the WorldCat library catalogue (subject searches were "San Francisco earthquake and fire 1906" and "Johnstown flood 1889") were done on March 26, 1998.

4. Paul Segall, "New Insights into Old Earthquakes," *Nature*, July 10, 1997, 122.

5. "100 Disaster Books," *NYT*, Apr. 24, 1906.

6. A search conducted on Apr. 14, 1997, of the 36 million records then available in the WorldCat library catalogue (subject search was "San Francisco earthquake and fire 1906") yielded 82 books, including imprints reissued by different publishers.

7. "Pernicious Advertising," *San Francisco Call*, May 20, 1906.

8. Quoted in William Issel and Robert W. Cherny, *San Francisco, 1865–1932: Politics, Power, and Urban Development* (Berkeley: Univ. of California Press, 1986), 39. For more on the economic history of the city, see pp. 23–52.

9. Santa Rosa, for example, was virtually leveled. See Robert L. Iacopi, *Earthquake Country*, 4th ed. (Tucson, Ariz.: Fisher Books, 1996), 96; Robert E. Wallace, ed., *San Andreas Fault System, California*, U.S. Geological Survey Professional Paper 1515 (Washington, D.C.: GPO, 1990), v, 159; and Dennis R. Dean, "The San Francisco Earthquake of 1906," *Annals of Science* 50 (1993): 510.

10. *San Francisco Relief Survey: The Organization and Methods of Relief Used After the Earthquake and Fire of April 18, 1906*, from studies compiled by Charles J. O'Connor et al. (New York: Survey Associates, 1913), 4. See also Judd Kahn, *Imperial San Francisco: Politics and Planning in an American City, 1897-1906* (Lincoln: Univ. of Nebraska Press, 1979), 130.

11. "Historical Clues from the San Andreas," *Science News*, Nov. 2, 1991, 286.

12. Jerry L. Coffman, Carl A. von Hake, and Carl W. Stover, eds., *Earthquake History of the United States* (Boulder, Colo.: Department of Commerce, NOAA, and Department of Interior, U.S. Geological Survey, 1982), 138; T.A. Heppenheimer, *The Coming Quake: Science and Trembling on the California Earthquake Frontier* (New York: Times Books, 1988), 76.

13. John McPhee, *Assembling California* (New York: Farrar, Straus and Giroux, 1993), 279.

14. Charles Wollenberg, "Life on the Seismic Frontier: The Great San Francisco Earthquake," *California History* 71 (Winter 1992/1993): 498. In an editorial attempting to play down the damage from the 1868 quake, the *Chronicle* wrote, "In every quarter of the city except that portion which rests upon made ground, magnificent buildings three and four stories in height are to be seen that show no traces of damage by the earthquake." See "The Exaggeration of Panic," *SFC*, Oct. 24, 1868.

15. See Karl V. Steinbrugge, *Earthquake Hazard in the San Francisco Bay Area: A Continuing Problem in Public Policy* (Berkeley: Institute of Governmental Studies, Univ. of California, 1968), 26.

16. The quoted material is from a letter Davidson sent to the Seismological Society of America in 1908. The letter is reprinted in William H. Prescott, "Circumstances Surrounding the Preparation and Suppression of a Report on the 1868 California Earthquake," *BSSA* 72 (1982): 2392. For biographical information on Gordon and Davidson, see Albert Shumate, *The California of George Gordon and the 1849 Sea Voyages of His California Association* (Glendale, Calif.: Arthur H. Clark, 1976); and Oscar Lewis, *George Davidson: A Pioneer West Coast Scientist* (Berkeley: Univ. of California Press, 1954).

17. J.D. Whitney, "Earthquakes," *North American Review* 108 (1869): 609. In 1906, a state commission formed to investigate the 1906 calamity also observed about the 1868 quake that a committee of scientists appointed to study this event found their report "supprest by the authorities, thru the fear that its publication would damage the reputation of the city." See Andrew C. Lawson, *The California Earthquake of April 18, 1906: Report of the State Earthquake Investigation Commission* (Washington, D.C.: Carnegie Institution, 1908), 1:434.

18. Marion G. Scheitlin, "Minimizing Effects of Earthquake in Frisco," *Insurance Field*, May 3, 1906, 17.

19. Quoted in Gladys Hansen and Emmet Condon, *Denial of Disaster: The Untold Story and Photographs of the San Francisco Earthquake and Fire of 1906* (San Francisco: Robert A. Cameron, 1989), 109.

20. Southern Pacific Company, *San Francisco Imperishable* (San Francisco: Southern Pacific Company, 1906), unpaginated. In *Sunset Magazine,* another publication established to promote business and tourism in California, the company railed against the so-called "calamity shriekers." Writing in the magazine in 1906, Charles Aiken, a prominent journalist, observed that of the $350 million in losses, only one-tenth could be blamed on the earthquake. "The quake did a lot of damage—no one questions that…," he wrote, "but that loss is insignificant when viewed beside the awful havoc of flames and dynamite." A little later in the year, Aiken reported on the city's rebuilding efforts: "Here are no signs of f[r]ight, nor panic, no indications of running away, forsooth, because the earth rocked a bit, and toppled a few walls, and broke water mains and let fierce fire work its bitter wrath." See Charles S. Aiken, "San Francisco's Plight and Prospect," *Sunset Magazine*, June/July 1906, 18, 19; and "San Francisco's Uprising," *Sunset Magazine*, Oct. 1906, 306.

21. This figure is from a report on the quake by the War Department. See Hansen and Condon, *Denial of Disaster*, 39.

22. Not surprisingly, you find very few close-up shots among the illustrations of the calamity pictured in *Sunset Magazine.* Most of the images show the city and its buildings from a distance. One panorama photograph taken from Nob Hill shortly after the fire began purportedly shows, says its caption, the "comparatively insignificant damage done by the earthquake." Readers are encouraged to "look at the picture and note the business section all undisturbed from any cause preceding the fire." One sees plenty of smoke in the background, while the foreground features skyscrapers, hotels, shops, and apartments "all standing as sturdily as the day they were built, ready for the business of the day should the flames spare them for it." The text accompanying the image admits that one can see "toppled chimneys innumerable, but all well-built structures of frame of steel or masonry withstood well the few seconds shock." What readers are not told is that the chosen camera angle could only reveal the most blatant effects of the quake. Needless to say, earthquake damage is often subtle in its visual manifestation but of devastating consequence nonetheless. See Aiken, "San Francisco's Plight and Prospect," 16–17, 19–20.

23. *Message of Governor George C. Pardee to the Extra Session of the Legislature of California*, June 2, 1906 (Sacramento: State Printing Office, 1906), 4, 10.

24. The report of the State Board of Trade is reprinted in "Faith in City Is Unshaken," *San Francisco Call*, May 27, 1906.

25. Quoted in "Quake Caused but Little Damage," *SFB*, May 15, 1906.

26. "The Future of the City," *SFC*, May 11, 1906. See also "Height of Buildings," *SFC*, June 20, 1906.

27. "Common Sense Prevailing over Hysterical Terror," *SFB*, Apr. 23, 1906. In another editorial, the paper wrote, "If the earth were quaking all the time, instead of shivering intermittently, as it does, the human race would quickly adapt itself to the movement and think no more about it....The terror of earthquakes lies in lack of familiarity. Had the earthquake of the 18th not been followed by the fire the damage done would not have been equal to that done by an ordinary tornado in the prairie States or by lightning last summer in New York city and its environs." See "What Would Happen If the Earth Fell to Quaking All the Time," *SFB*, May 10, 1906.

28. The report on the meeting is discussed in "Big Structures Now Planned," *SFC*, Apr. 25, 1906.

29. Newlands is famous for proposing the so-called Newlands Act of 1902, which took money from the sale of western lands and placed it in an irrigation fund used to reclaim more desert.

30. Francis G. Newlands, "The New San Francisco," *Independent*, May 10, 1906, 1093, 1094, 1095.

31. My thinking here was influenced by Mary Douglas and Aaron Wildavsky, *Risk and Culture: An Essay on the Selection of Technological and Environmental Dangers* (Berkeley: Univ. of California Press, 1982), 29–30.

32. Gilbert continued: "Pains are taken to speak of the disaster of 1906 as a conflagration, and so far as possible the fact is ignored that the conflagration was caused, and its extinguishment prevented, by injuries due to the earthquake. During the period of after-shocks, it was the common practise of the San Francisco dailies to publish telegraphic accounts of small tremors perceived in the eastern part of the United States, but omit mention of stronger shocks in the city itself." G.K. Gilbert, "Earthquake Forecasts," *Science*, Jan. 22, 1909, 135.

33. J.C. Branner, "Earthquakes and Structural Engineering," *BSSA* 3 (1913): 2–3.

34. Andrew C. Lawson, "Seismology in the United States," *BSSA* 1 (1911): 3.

35. In addition to Gilbert, Branner, and Lawson, Harry Fielding Reid of Johns Hopkins and four other scientists were appointed to the commission.

36. The Lawson *California Earthquake* report was published in two volumes; volume one in 1908 and volume two in 1910. "No report of any previous earthquake has been issued on so liberal a scale," wrote Charles Davison some years later in his study of the early history of seismology. Charles Davison, *The Founders of Seismology* (Cambridge: Cambridge Univ. Press, 1927), 152–53.

37. "Adjustments Are Now at a Halt," *Insurance Field*, May 24, 1906, 6. California's insurance commissioner E. Myron Wolf tried to drive home this point by inviting

the presidents of all fire insurance companies to come to San Francisco to see for themselves the minimal damage the earthquake had caused. In Wolf's opinion, that damage amounted to no more than 2 to 3 percent of the total destruction. But the companies themselves were more interested in setting up so-called dope departments to counter the massive propaganda campaign being waged by both the business class in league with policyholders throughout the city than in going on a disaster junket. These departments went in search of photographs taken on the morning of the earthquake before the fire consumed the city. Many such photographs were shot by amateurs seeking to memorialize the historic event. Unsurprisingly, such photographs were in short supply, especially in the months immediately after the calamity. See "Reliable Data Collected," *Insurance Field*, July 5, 1906, 11.

38. Frederick L. Hoffman, *Earthquake Hazards and Insurance* (Chicago: Spectator Co., 1928), 128. In an effort to come to some kind of compromise on the adjustment question, representatives of the fire insurance companies met in Oakland in June following the calamity. A proposal was made to settle claims at 75 percent of the face value of each policy in order to account for the hard-to-discern losses caused by the earthquake but often masked by the fire. Over 60 companies—the "six-bit" firms—voted in favor of this quarter reduction, often referred to as the horizontal cut. Thirty-two companies—the "dollar-for-dollar" firms—voted against the resolution and chose to settle losses in accordance with the full terms and conditions stipulated in their policies. Almost half of the gross insurance loss was settled by these dollar companies.

At least 14 companies doing business in the city had inserted so-called earthquake clauses in their insurance contracts. For example, the clause contained in policies written by the Williamsburgh City Fire Insurance Company of Brooklyn, New York, exempted the company from "loss caused directly or indirectly by invasion…or for loss or damage occasioned by or through any volcano, earthquake, or hurricane." After the calamity, Williamsburgh City invoked the clause to totally deny all liability for damage. Policyholders, in turn, sued the company. In the leading case of *Williamsburgh City Fire Ins. Co. of Brooklyn v. Willard*, 164 F. 404 (9th Cir. 1908), a federal circuit court of appeals interpreted the earthquake clause narrowly, forcing the company to pay the policyholder claims. Williamsburgh had argued that since the fire that destroyed the property in question could be traced back to the earthquake, the company was exempt from any liability for damage. The court, however, saw things differently. When an earthquake was the direct and proximate cause of the damage in question, the company remained exempt from paying losses. But when an earthquake—like the one in 1906— indirectly caused damage, the company paid.

39. "California Not an Earthquake Country," *Coast Review*, Sept. 1906, pp. g, h.

40. See "The Effects of the San Francisco Earthquake of April 18th, 1906, on Engineering Construction," *Transactions of the American Society of Civil Engineers*, 59 (1907): 208–209, 211.

41. *Ordinances of the City and County of San Francisco*, Building Law, Ordinance No. 31, § 69, approved July 5, 1906.

42. *The Building Law and the Plumbing Law of the City and County of San Francisco,* Bill No. 1121, Ordinance No. 1008 (New Series), § 89, approved Dec. 22, 1909.

43. *Official Building Laws, City and County of San Francisco* (San Francisco: Daily Pacific Builder, 1921), § 89.

44. Bailey Willis, "Earthquake Risk in California," *BSSA* 14 (1924): 15.

45. John Ripley Freeman, *Earthquake Damage and Earthquake Insurance* (New York: McGraw Hill, 1932), 30.

46. Carl-Henry Geschwind, "Earthquakes and Their Interpretation: The Campaign for Seismic Safety in California, 1906–1933" (Ph.D. diss., Johns Hopkins Univ., 1996), 155, 156–157, 159.

47. Freeman, *Earthquake Damage,* 693–94.

48. Ibid., 11. "Previous to 1925 the demand for earthquake insurance on the Pacific Coast was rather small," explained the engineer H.M. Engle of the Board of Fire Underwriters of the Pacific. "The severe shock at Santa Barbara in that year, with the extensive damage produced, was a startling reminder to the public at large that the earthquake hazard on the Coast is not a thing of the past and that people may be justified in looking forward to the future with some apprehension." H.M. Engle, "The Earthquake Resistance of Buildings from the Underwriters' Point of View," *BSSA* 19 (1929): 86.

49. Coffman, von Hake, and Stover, *Earthquake History,* 10.

50. Arthur Pound, "Conquering the Earthquake Crisis," *Independent,* July 25, 1925, 95.

51. In 1930, premiums on California properties alone were more than $2 million. See Freeman, *Earthquake Damage,* 658.

52. See C.T. Manwaring, "Report of Committee on Building For Safety Against Earthquakes: Preliminary Report on Guarding Against Panic," *BSSA* 15 (1925): 213–221.

53. Issel and Cherny, *San Francisco, 1865–1932,* 42.

54. See, e.g., Stuart Ewen, *PR!: A Social History of Spin* (New York: Basic Books, 1996), 131–32. Still, recognition of the earthquake hazard came slowly. When John Freeman sat down in the early 1930s to write *Earthquake Damage and Earthquake Insurance,* the most exhaustive effort to date exploring the extent of seismic damage in the United States, California was still mired in the throes of denial. "Neither state nor municipal government in California," he observed, "has given evidence of any strong interest in the scientific study of earthquakes." The government funds that existed to study quakes, he noted, came mostly from "Eastern sources." Freeman continued, "There seems to have been a local attempt to suppress information about earthquakes, lest it hurt California business, and rumors of this suppression have reacted unfavorably, by increasing the apprehensions of Eastern underwriters." To be sure, there were some signs of progress. San Francisco had reorganized its Bureau of Building Inspection and had appointed a reputable structural engineer to be superintendent. But by and large, authorities in San Francisco, and California more generally, had made little progress on seismic preparedness in the years after the 1906 calamity. As Freeman wrote, "the pos-

sibilities of safeguarding human life and property against another earthquake by means of proper structural designs, by improved building laws, by more rigorous inspection and by removing existing hazards of weak parapets, etc., were strangely forgotten." San Francisco did pass an ordinance outlawing parapets after the 1906 disaster, but as late as 1989 had never properly enforced the law. Freeman, *Earthquake Damage*, 2, 658, 700.

55. "Murder in California," *New Republic*, Mar. 22, 1933, 146.

56. Geschwind, "Earthquakes," 214.

57. Ibid., 228–231. Geschwind's dissertation argues against the views set forth in Arnold J. Meltsner, "The Communication of Scientific Information to the Wider Public: The Case of Seismology in California," *Minerva* 17 (1979): 331–354. Meltsner claims that scientific knowledge about California's earthquake risk was suppressed by business interests. Geschwind shows, however, that at the time of the 1906 earthquake, scientists themselves were ignorant of the region's seismic risk. In short, there was nothing, he explains, for the business community *to* suppress. Although there is clearly some real merit to his critique, Geschwind tends to privilege science over popular knowledge. Clearly, people were familiar with the region's vulnerability to earthquakes since at least the 1860s, even if scientists had yet to uncover the critical details behind these disturbances. Also, it is important to note that the scientists themselves discussed attempts to suppress the science behind earthquakes, as we have seen, an issue Geschwind overlooks.

58. *Fireman's Fund Record*, Apr. 1936, 6, 7.

59. "The Candid Cameraman," *Esquire*, Sept. 1936, 98.

60. "Seismic Note," *NYT*, July 12, 1936.

61. "Cinema," *Time*, July 6, 1936, 48.

62. Wagner's review is reprinted in Anthony Slide, ed., *Selected Film Criticism 1931–1940* (Metuchen, N.J.: Scarecrow Press, 1982), 217.

63. Charles W. Jennings, "New Geologic Map of California: A Summation of 140 Years of Geologic Mapping," *California Geology* 31 (1978): 77. On the fault map published by the Seismological Society in 1923, see Geschwind, "Earthquakes," 136–141.

64. "The Earthquake Taboo," *Nation*, Nov. 30, 1964, 405.

65. John J. Fried, *Life Along the San Andreas Fault* (New York: Saturday Review Press, 1973), 124–126.

66. "The Case of the Muzzled Geologists," *SFC*, Jan. 24, 1965, magazine sec., 7; Fried, *San Andreas*, 128–130.

67. Fried, *San Andreas*, 133–136.

68. Karl V. Steinbrugge, *Earthquakes, Volcanoes, and Tsunamis: An Anatomy of Hazards* (New York: Skandia America Group, 1982), 32.

69. According to one estimate, California and western Nevada account for roughly 90 percent of all seismic activity in the conterminous United States. See Coffman, von Hake, and Stover, *Earthquake History*, 137.

70. "Seismic Study Shakes Up Utah," *Engineering News-Record*, Nov. 30, 1992, 24.

71. Steinbrugge, *Earthquakes, Volcanoes*, 3.

72. McPhee, *Assembling California*, 273.

73. Steinbrugge, *Earthquake Hazard*, 14.
74. See "Maps That Help to Spot Where Trouble Lies in Store," *New Scientist*, May 12, 1990, 58.
75. Quoted in Fried, *San Andreas*, 10.
76. Karl V. Steinbrugge et al., *Earthquake Planning Scenario for a Magnitude 7.5 Earthquake on the Hayward Fault in the San Francisco Bay Area* (Sacramento: California Department of Conservation, Division of Mines and Geology, 1987).
77. Risa Palm and Michael E. Hodgson, *After a California Earthquake: Attitude and Behavior Change* (Chicago: Univ. of Chicago Press, 1992), 92.
78. *Who Perished: A List of Persons Who Died as a Result of the Great Earthquake and Fire in San Francisco on April 18, 1906*, comp. Gladys Hansen (San Francisco: San Francisco Archives, 1980), 3.
79. Thurston Clarke, *California Fault: Searching for the Spirit of a State Along the San Andreas* (New York: Ballantine, 1996), 205–207. According to Clarke, Hansen predicts a final death toll of somewhere between 5,000 and 10,000.
80. Hansen and Condon, *Denial of Disaster*, 18, 20, 23.
81. McPhee, *Assembling California*, 301–302.
82. Mary C. Comerio, *Disaster Hits Home: New Policy for Urban Housing Recovery* (Berkeley: Univ. of California Press, 1998), 67, table 12; 73; 81.
83. Steinbrugge, *Earthquakes, Volcanoes*, 5.
84. Quoted in "Quake Improvements Put a Price on Life," *SFC*, Jan. 22, 1994.
85. Quoted in "Battle over Cost of Quake Retrofitting," *SFC*, Feb. 6, 1995.

THREE · DO-IT-YOURSELF DEATHSCAPE

1. David W. Lantis, Rodney Steiner, and Arthur E. Karinen, *California: Land of Contrast* (Belmont, Calif.: Wadsworth, 1963), 308, fig. 8.25.
2. Polly Redford, *Billion-Dollar Sandbar: A Biography of Miami Beach* (New York: Dutton, 1970), 48, quotation from p. 45.
3. Luther J. Carter, *The Florida Experience: Land and Water Policy in a Growth State* (Baltimore: Johns Hopkins Univ. Press, 1974), 74. Fisher bought the property from John Collins, a New Jersey investor who used the cash to finish building a bridge to his remaining property.
4. Ruby Leach Carson, "Forty Years of Miami Beach," *Tequesta* 15 (1955): 11.
5. Redford, *Billion-Dollar Sandbar*, 73.
6. Frank Parker Stockbridge and John Holliday Perry, *Florida in the Making* (Jacksonville, Fla.: de Bower Publishing Co., 1926), 289.
7. Redford, *Billion-Dollar Sandbar*, 154.
8. This discussion of island building is based on Redford, *Billion-Dollar Sandbar*, 99; Carter, *Florida Experience*, 76; Carson, "Forty Years of Miami Beach," 14, 19; and Millicent Todd Bingham, "Miami: A Study in Urban Geography," *Tequesta* 8 (1948): 99.
9. One might rightly wonder how the bottom of Biscayne Bay—a navigable waterway and therefore under the common law supposedly held in trust for the public

by the state—wound up in private hands. Even Florida's very own supreme court recognized as late as 1909 that the disposal of a vast amount of bottom land in a navigable waterway would violate the so-called public trust doctrine. Although such land was supposed to be held in trust for the benefit of the state's citizens, this doctrine's common-law roots, it turns out, meant that it could be modified by legislative decree. See Frank E. Maloney, Sheldon J. Plager, and Fletcher N. Baldwin Jr., *Water Law and Administration: The Florida Experience* (Gainesville: Univ. of Florida Press, 1968), 356.

10. John M. De Grove, "Administrative Problems in the Management of Florida's Tidal Lands," in Per Bruun and John M. De Grove, *Bayfill and Bulkhead Line Problems—Engineering and Management Considerations*, Studies in Public Administration, no. 18 (Gainesville: Public Administration Clearing Service, Univ. of Florida, 1959), 20–22.

11. Carson, "Forty Years of Miami Beach," 20.

12. Redford, *Billion-Dollar Sandbar*, 107, 154.

13. Ralph Middleton Munroe and Vincent Gilpin, *The Commodore's Story* (n.p.: Ives Washburn, 1930), 340–342.

14. John Kenneth Galbraith, *The Great Crash 1929* (1955; reprint, New York: Avon, 1979), 3.

15. "In the Wake of the Hurricane," *Saturday Evening Post*, Nov. 27, 1926, 60; Frank Bowman Sessa, "Real Estate Expansion and Boom in Miami and Its Environs During the 1920s" (Ph.D. diss., Univ. of Pittsburgh, 1950), 332.

16. "Catastrophe Is Climax to Economic Disaster in Florida," *New York Evening Post*, Sept. 20, 1926.

17. "Causeway Traffic Delayed by Storm," *MH*, July 27, 1926.

18. Editorial, *MH*, July 28, 1926.

19. "City Well Protected," *MH*, July 31, 1926.

20. Quotations are from "Miamian Grinds Camera All Through Hurricane to Record Scene for Movies," *Tampa (Fla.) Tribune*, Sept. 22, 1926.

21. "The Row over Florida Relief," *Literary Digest*, Oct. 16, 1926, 17.

22. This figure is derived from table 2 in Paul J. Hebert, Jerry D. Jarrell, and Max Mayfield, *The Deadliest, Costliest, and Most Intense United States Hurricanes of This Century (And Other Frequently Requested Hurricane Facts)*, NOAA Technical Memorandum NWS TPC-1 (Washington, D.C.: Department of Commerce, 1996), 6.

23. For the 11 years between 1924 and 1934, windstorm losses in Florida totaled more than $23 million. According to F.W. Brundig, the manager of the Florida Underwriters Agency, "with an average loss ratio of 179 percent for the period 1924–1934, which will be materially increased when the 1935 hurricane losses are added, you can appreciate why all companies look askance at windstorm liability." Quoted in "Florida Windstorm in Red," *National Underwriter*, Mar. 26, 1936, 10.

24. "Normalcy," *MH*, Sept. 30, 1926.

25. Nixon Smiley, *Knights of the Fourth Estate: The Story of the* Miami Herald (Miami: E.A. Seeman, 1974), 27, 29–30, 81.

26. "Ninety Days Hence," *MH*, Sept. 25, 1926. The year following the Miami hurricane, the press in the city of New Orleans adopted a similar strategy of denying disaster in an effort to head off unfavorable publicity surrounding the monstrous 1927 Mississippi River flood. See John M. Barry, *Rising Tide: The Great Mississippi Flood of 1927 and How It Changed America* (New York: Simon and Schuster, 1997), 225–227, 239–240.

27. Quoted in "Peay and God," *Tampa Tribune*, Sept. 25, 1926.

28. "Wind and Wickedness," *MH*, Sept. 29, 1926. It was, of course, not uncommon in the early twentieth century for those giving sermons following disasters to take the opportunity to warn that the destruction was connected in some way to America's emerging consumer culture. See, e.g., Steven Biel, *Down with the Old Canoe: A Cultural History of the Titanic Disaster* (New York: Norton, 1996), 64.

29. "Proclamation," *MH*, Sept. 20, 1926.

30. "Do Not Believe Rumors," *MH*, Sept. 21, 1926.

31. "In Today's News," *MH*, Sept. 26, 1926. One of the more "exaggerated" accounts, at least in terms of the number of deaths reported, appeared in the *San Francisco Examiner*. The paper reported "approximately 1,215 persons" killed in the storm. The *Examiner*, which had sought to downplay the destruction of the earthquake of two decades earlier, was willing to accept reports of abject destruction elsewhere in the interest of selling papers. "1215 Dead, 2000 Hurt as Gale Razes Florida Cities," *SFE*, Sept. 20, 1926.

32. "Miamians Criticize Martin," *NYT*, Oct. 2, 1926.

33. "Fury of Hurricane Told by Witnesses," *NYT*, Sept. 21, 1926; "Modern Buildings Playthings of Wind, Eyewitness Says," *Washington Post*, Sept. 21, 1926.

34. "To the American People," *MH*, Sept. 23, 1926.

35. Quoted in "The Row over Florida Relief."

36. Quoted in "Reports Floridians Admit Need of Aid," *NYT*, Oct. 4, 1926. See also "Red Cross Accuses Florida's Governor," *NYT*, Oct. 2, 1926.

37. "Governor Refuses Special Session for Relief Work," *Homestead (Fla.) Enterprise*, Oct. 1, 1926.

38. Quoted in "Florida Damage of Local Import," *Wall Street Journal*, Oct. 8, 1926.

39. Ibid.

40. Quoted in "The Row over Florida Relief."

41. Quoted in "Red Cross Accuses Florida's Governor." For more on Payne's response to the state's actions, see "President Praises Red Cross as Ideal," *NYT*, Oct. 5, 1926.

42. Quoted in "The Row over Florida Relief."

43. "In the Wake of the Hurricane," 59.

44. "Will South Florida Wake Up?" *Homestead Enterprise*, Oct. 8, 1926.

45. Joe Hugh Reese, *Florida's Great Hurricane* (Miami: Lysle E. Fesler, 1926), 53, 54, 57. Reese received his information that 300 people had died from Howard Sharp, editor of the *Everglades News*.

46. Carter, *Florida Experience*, 68–69, 71, 73.

47. Alfred Jackson Hanna and Kathryn Abbey Hanna, *Lake Okeechobee: Wellspring of the Everglades* (Indianapolis: Bobbs-Merrill, 1948), 254–255.

48. Quoted in "Lower the Lake," *Tampa Tribune*, Sept. 28, 1926.
49. See "Day of Death Follows Night of Horror" and "The Dead Accuse," *Everglades News* (Canal Point, Fla.), Sept. 24, 1926.
50. On land speculation companies and the Everglades, see "They Can Get What They Want," ibid., Oct. 29, 1926.
51. Johnson's statement is reprinted in "Apology for the Drainage Board in Press Story," *Homestead Enterprise*, Oct. 22, 1926.
52. Quoted in "Martin Pleads Lack of Funds," *MDN*, Oct. 28, 1926.
53. "The Wind and Executive Decree," *Everglades News*, Nov. 12, 1926. A device for measuring wind speed on the lake had been broken in the July 1926 hurricane.
54. Although the Red Cross estimated 1,770 deaths, the most commonly quoted figure on the number dead is 1,836. See American National Red Cross, *The West Indies Hurricane Disaster, September, 1928: Official Report of Relief Work in Porto Rico, the Virgin Islands and Florida* (Washington, D.C.: American Red Cross, 1929), 53; and table 2 in Hebert, Jarrell, and Mayfield, *Deadliest*, 6.

    The Weather Bureau did not predict the 1928 hurricane until the night before it occurred; the Seminole Indians, however, interpreted an unseasonable blossoming of saw grass as a sign of an impending hurricane. Although the Indians may have been on the mark with their prediction in 1928, as Andrew Ross notes, there were other occasions when their predictions proved inferior to those of the Weather Bureau. See Andrew Ross, *Strange Weather: Culture, Science, and Technology in the Age of Limits* (London: Verso, 1991), 216. For a fictional account of the 1928 hurricane, see Zora Neale Hurston, *Their Eyes Were Watching God* (1937; reprint, New York: Harper and Row, 1990), 147ff.
55. Quoted in Lawrence E. Will, *Okeechobee Hurricane and the Hoover Dike* (St. Petersburg, Fla.: Great Outdoors Publishing, 1961), 84.
56. Ibid., 88, quotation from pp. 89–90.
57. "A Measurable Disaster," *Wall Street Journal*, Sept. 19, 1928.
58. "A Touch of Hysteria," ibid., Sept. 22, 1928.
59. Quoted in Will, *Okeechobee Hurricane*, 83. In an attempt to make sure that those outside Florida did not view the 1928 hurricane as a "Florida disaster," the Clearwater Chamber of Commerce sent out 5,000 letters explaining that, in the words of the *Clearwater Sun and Herald*, it was "almost impossible for such storms to affect the Clearwater and West Coast sections" of the state. "5,000 Letters Explaining Immunity of West Coast from Serious Hurricane Damage Sent Out by C. C.," *Clearwater (Fla.) Sun and Herald*, Sept. 30, 1928.
60. "Lake Rise of 3 Feet in 30 Days Can Repeat," *Belle Glade News* (Canal Point, Fla.), Sept. 7, 1928.
61. "Defend Elliot Say Flood Was 'An Act of God,'" *Everglades News*, Oct. 19, 1928. See also "Terrible Responsibility," ibid., Oct. 26, 1928; and "An Act of God?" ibid., Nov. 9, 1928.
62. Quoted in Will, *Okeechobee Hurricane*, 103.
63. The barometric pressure registered 26.35 inches, making it the lowest recorded pressure of a storm making landfall in the United States. Hurricane Gilbert in

1988, however, is actually the most intense storm on record, with a slightly lower reading of 26.23 inches. See table 4 in Hebert, Jarrell, and Mayfield, *Deadliest*, 9; and Roger A. Pielke Jr., "Reframing the U.S. Hurricane Problem," *Society and Natural Resources* 10 (1997): 491.

64. "Naming the Hurricane," *MH*, Sept. 14, 1935.

65. Gary Dean Best, *FDR and the Bonus Marchers, 1933–1935* (Westport, Conn.: Praeger, 1992), 5.

66. See, e.g., "The Whys of Matecumbe," *Daily Tribune* (Miami), Sept. 5, 1935; and "What Price Buck Passing?" ibid., Sept. 6, 1935.

67. Quoted in "Storm Deaths An Act of God Says Williams," *MDN*, Sept. 9, 1935, which reprinted the report done by Williams and Ijams.

68. Greater Miami Ministerial Association to Franklin D. Roosevelt, Sept. 10, 1935, Records of the Works Progress Administration, Federal Emergency Relief Administration, Central Files, Record Group 69.006.1, National Archives, Washington, D.C.

69. Frank A. Hamilton to Franklin D. Roosevelt, Sept. 11, 1935, ibid.

70. George W. Miller to Harry L. Hopkins, Sept. 27, 1935, ibid.

71. This description of the railroad project is based on Oliver Griswold, *The Florida Keys and the Coral Reef* (Miami: Graywood Press, 1965), 58–59; and Carter, *Florida Experience*, 66.

72. Griswold, *Florida Keys*, 60–61, quotation from p. 61.

73. Ibid., 59, 69. For more on the bottling up of water caused by the destruction of the natural outlets, see "Matecumbe Investigation," *Daily Tribune*, Sept. 7, 1935.

74. Transcript in investigation "Re Hurricane Florida," Records of the Works Progress Administration, Federal Emergency Relief Administration, Central Files, Record Group 69.006.1, National Archives, 870–880, quotations from pp. 878, 879.

75. House Committee on World War Veterans' Legislation, *Florida Hurricane Disaster*, 74th Cong., 2d sess., 1935, Committee Print, 30, 31. How Wilcox arrived at the figure of 4,000 for the number of Floridians killed in hurricanes since 1925 is unclear, though it could be accurate.

76. Peter J. May, *Recovering from Catastrophes: Federal Disaster Relief Policy and Politics* (Westport, Conn.: Greenwood, 1985), 27.

77. Just two months after the September 1935 disaster, another hurricane descended on Miami. Only a handful of people died, although several millions of dollars in damage was done. The *Miami Daily News*, not always as fierce in its progrowth proclivities as the *Herald*, but a city booster nonetheless, declared that Miami had withstood the storm's test. More people die in traffic accidents in a single day than died in the storm, it pointed out. As the paper continued, "the chief suffering from any hurricane is caused by hysteria." "Miami Stands the Test," *MDN*, Nov. 5, 1935.

78. "Naming the Hurricane."

79. George R. Stewart, *Storm* (New York: Random House, 1941), 12.

80. "Why Gales Are Gals," *NYT Magazine*, Sept. 26, 1954, 57.

81. "Vicious Lady," *Time*, Sept. 5, 1949, 16. "From now on, the U.S. Weather Bureau will distinguish Atlantic hurricanes by girls' names, as the Air Force and Navy

have been doing for years in the Pacific," wrote *Time* in 1953. See "Alice to Wallis," *Time*, June 15, 1953; and Jay Barnes, *Florida's Hurricane History* (Chapel Hill: Univ. of North Carolina Press, 1998), 188, 195.

The feminization of hurricanes is interesting in that it serves as proof of what anthropologist Sherry Ortner has called the "mutual metaphorization" that occurs between gender and the nature/culture dichotomy. The natural world sets limits on human action, and it is at precisely the point where culture collides with nature in its more uncontrolled state that gender was employed as a mediating language. See Sherry B. Ortner, *Making Gender: The Politics and Erotics of Culture* (Boston: Beacon, 1996), 179. In 1978, under pressure from the National Organization for Women and other groups, the Department of Commerce announced that it would name hurricanes after men as well. See "Another Sexist Bastion Falls," *NYT*, May 13, 1978.

82. Elaine Tyler May, *Homeward Bound: American Families in the Cold War Era* (New York: Basic Books, 1988), 99, 109–110.

INTERLUDE · BODY COUNTING

1. Figures on the number of people killed in natural disasters only became reliable in the twentieth century. Still, the fact that a number of massively deadly disasters occurred during the period between 1880 and 1930 makes it certain that this period ranks as the deadliest in U.S. history.

2. See James Lal Penick Jr., *The New Madrid Earthquakes*, rev. ed. (Columbia: Univ. of Missouri Press, 1981), 109. Although the New Madrid earthquakes occurred before the development of the Richter scale, seismologists believe that the disturbances had the largest magnitudes and affected the greatest area of any shock in historic time. See ibid., 141, 147; and Bruce A. Bolt, *Earthquakes and Geological Discovery* (New York: Scientific American Library, 1993), 118.

3. David M. Ludlum, *Early American Hurricanes, 1492–1870* (Boston: American Meteorological Society, 1963), 77, 80.

4. Although tornadoes posed few problems before the late nineteenth century, there were exceptions, including the Great Natchez tornado of 1840, which killed upwards of 300, and the 1860 Camanche tornado, which struck eastern Iowa and killed 141. See David M. Ludlum, *Early American Tornadoes, 1586–1870* (Boston: American Meteorological Society, 1970), 86, 131.

5. See table 2 in Paul J. Hebert, Jerry D. Jarrell, and Max Mayfield, *The Deadliest, Costliest, and Most Intense United States Hurricanes of This Century (And Other Frequently Requested Hurricane Facts)*, NOAA Technical Memorandum NWS TPC-1 (Washington, D.C.: Department of Commerce, 1996), 6. Although it is difficult to say precisely how many died, the hurricane did its greatest damage on the Sea Islands, inhabited by poor blacks.

6. "The Cyclone in the South," *Harper's Weekly*, Sept. 16, 1893, 882; "The Sea Island Hurricanes," *Scribner's Magazine*, Feb. 1894, 240. Coming just seven years after the 1886 earthquake, the hurricane reignited the debate over the effect of relief on

work-discipline. In a letter to the *News and Courier*, one observer noted that distributing food and money to blacks on the Sea Islands was likely to do "more harm than good." Drawing an analogy between life for blacks on the islands and those who lived along the floodplains of the Mississippi and other rivers, the author noted that providing such people with relief aid has always "led to demoralization and promoted idleness." See "Work Provided Is Real Charity" and "Help Them to Help Themselves," *NC*, Sept. 6, 1893.

7. See "The Cyclone Sufferers," *Evening Post* (New York), Oct. 7, 1893. In New Orleans, the press sought to walk a very fine line between acknowledging the incredible magnitude of the calamity and denying that a storm of similar strength would visit the area again in the near future. Alluding to the earlier hurricane along the coasts of Georgia and South Carolina, the *Daily Picayune* noted that the year 1893 would go down as "a memorable period in storm annals in this section." Memorable for what was the question. The paper was vehement that the memory of the disaster not be allowed to interfere with the further economic development of the New Orleans region. "With a railroad nearly to Grand Isle, and a short ferry to the mainland, the place will be more accessible, and, if it shall be put in proper condition, the recollection of the storm of 1893 will soon become a sad and solemn memory, and not a bugbear to frighten people away from their enjoyment." See "The Recent Storm Compared with the Last Island Disaster," *Daily Picayune* (New Orleans), Oct. 5, 1893; and "A Region Not to be Abandoned," ibid., Oct. 14, 1893.

8. See table 2 in Hebert, Jarrell, and Mayfield, *Deadliest*, 6.

9. The classic text on the flood is David McCullough, *The Johnstown Flood: The Incredible Story Behind One of the Most Devastating "Natural" Disasters America Has Ever Known* (New York: Simon and Schuster, 1968). The flood occurred when heavy rains caused an earthen dam to break. The dam had been built at the request of a wealthy sporting club, patronized by Andrew Carnegie and Henry Clay Frick, among others, in order to create a lake for recreational purposes. However, it appears that the dam was of poor construction. Not surprisingly perhaps, the club and its lawyers claimed that the flood, which roared through the Conemaugh Valley, was a "visitation of providence." The club's resident engineer, John Parke, told the *New York Sun* that, "no blame can be attached to anyone for this greatest of horrors. It was a calamity that could not be avoided." Some of the victims, however, took a contrary view. George Swank, a local newspaper editor, wrote, "We think we know what struck us, and it was not the hand of Providence. Our misery is the work of man." Quoted in ibid., 214, 253.

10. J.W. Powell, "Our Recent Floods," *North American Review* 155 (1892): 149.

11. See the table in David M. Ludlum, *The American Weather Book* (Boston: Houghton Mifflin, 1982), 80.

12. John M. Barry, *Rising Tide: The Great Mississippi Flood of 1927 and How It Changed America* (New York: Simon and Schuster, 1997), 158. Ludlum cites a figure of 467 deaths in the Ohio floods of 1913. See the table in Ludlum, *American Weather Book*, 80.

13. See table 3.7, "National Weather Service Estimates of Flood-Related Deaths in the United States, 1916–1989" in *Floodplain Management in the United States: An Assessment Report*, prepared for the Federal Interagency Floodplain Management Task Force by L. R. Johnston Associates (1992), 2: 3–17.

14. Quoted in Barry, *Rising Tide*, 286.

15. Ibid. The NWS estimated 423 flood deaths in 1927. See table 3.7, "National Weather Service Estimates." Another estimate places the death toll between 250 and 500. See Pete Daniel, *Deep'n As It Come: The 1927 Mississippi River Flood* (New York: Oxford Univ. Press, 1977), 10.

16. The dates and number of deaths in the 10 most deadly tornado outbreaks are as follows: Mar. 18, 1925 (747); Mar. 21–22, 1932 (330); Apr. 23–24, 1908 (324); Apr. 3–4, 1974 (307); May 27, 1896 (305); Apr. 11, 1965 (256); Apr. 5, 1936 (249); Apr. 20, 1920 (224); May 8–9, 1927 (217); Apr. 6, 1936 (205). See table 4.5 in Thomas P. Grazulis, *Significant Tornadoes, 1680–1991* (St. Johnsbury, Vt.: Environmental Films, 1993), 38.

17. "The Deadliest Tornado in American History," *Literary Digest*, Apr. 4, 1925, 48. For more on the 1925 tornado outbreak, see Jan Brodt, "The Tri-State Tornado, March 18, 1925," *Weatherwise* 39 (1986): 91–94. No other tornado outbreak even comes close to rivaling the 1925 event, save perhaps for a swarm of tornadoes that began in Mississippi in February 1884 and plowed through the South. Estimates of the number dead range from a low of 167 to a high of 1,200. See table 4.5 in Grazulis, *Significant Tornadoes*, 38; and Joseph G. Galway, "Ten Famous Tornado Outbreaks," *Weatherwise* 34 (1981): 102. The latter figure, if true, would of course make it far and away the most deadly tornado disaster ever to happen in this country. Weather historian David Ludlum, who argues for a death toll of roughly 800, has observed that the calamity has failed to establish any "lasting notoriety," in part, he speculates, because the victims were largely poor blacks living in isolated areas, a fact that might also suggest that the actual number of deaths has been underreported. See David M. Ludlum, "The Great Tornado Outbreak on 19 February 1884," *Weatherwise* 28 (1975): 85.

18. See fig. 1.1 in Grazulis, *Significant Tornadoes*, 2.

19. The tables on page 227 describe some key population trends in turn-of-the-century America. The tables are derived from the following sources: Department of Commerce, U.S. Bureau of the Census, *Thirteenth Census of the United States: Population 1910: General Report and Analysis* (Washington, D.C.: GPO, 1913), 1:87, 88, 96; and *Census of Population: 1950, United States Summary* (Washington, D.C.: GPO, 1953), 2, pt. 1: 46–47.

20. These figures are derived from table 2 in Fred Doehring, Iver W. Duedall, and John M. Williams, *Florida Hurricanes and Tropical Storms: 1871–1993: An Historical Survey*, Technical Paper no. 71 (Gainesville: Florida Sea Grant College Program, 1994), 53.

21. Grazulis, *Significant Tornadoes*, 196, 198.

22. Quoted in Kai T. Erikson, *Everything in Its Path: Destruction of Community in the Buffalo Creek Flood* (New York: Simon and Schuster, 1976), 178.

Population Growth in Cities of Extreme Hurricane Risk, 1880-1910

City	1880	1890	1900	1910	% Change, 1880–1910
Baton Rouge, La.	7,197	10,478	11,269	14,897	107
Brownsville, Tex.	—	6,134	6,305	10,517	—
Corpus Christi, Tex.	3,257	4,387	4,703	8,222	152
Galveston, Tex.	22,248	29,084	37,789	36,981	66
Houston, Tex.	16,513	27,557	44,633	78,800	377
Mobile, Ala.	29,132	31,076	38,469	51,521	77
New Orleans, La.	216,090	242,039	287,104	339,075	57
Pensacola, Fla.	—	11,750	17,747	22,982	—

Population Growth in Cities of Extreme Seismic Risk, 1880-1910

City	1880	1890	1900	1910	% Change, 1880–1910
Berkeley, Calif.	—	5,101	13,214	40,434	—
Oakland, Calif.	34,555	48,682	66,960	150,174	335
San Francisco, Calif.	233,959	298,997	342,782	416,912	78
Long Beach, Calif.	—	564	2,252	17,809	—
Los Angeles, Calif.	11,183	50,395	102,479	319,198	2,754
Pasadena, Calif.	—	4,882	9,117	30,291	—
Memphis, Tenn.	33,592	64,495	102,320	131,105	290
St. Louis, Mo.	350,518	451,770	575,238	687,029	96

Population Growth in Cities of Moderate to Extreme Tornado Risk, 1880-1910

City	1880	1890	1900	1910	% Change, 1880–1910
Cincinnati, Ohio	255,139	296,908	325,902	363,591	43
Des Moines, Iowa	22,408	50,093	62,139	86,368	285
Indianapolis, Ind.	75,056	105,436	169,164	233,650	211
Kansas City, Mo.	55,785	132,716	163,752	248,381	345
Kansas City, Kans.	3,200	38,316	57,418	82,331	2,473
Louisville, Ky.	123,758	161,129	204,731	223,928	81
Memphis, Tenn.	33,592	64,495	102,320	131,105	290
Minneapolis, Minn.	46,887	164,738	202,718	301,408	543
Nashville, Tenn.	43,350	76,168	80,865	110,364	155
Omaha, Nebr.	30,518	140,452	102,555	124,096	307
Peoria, Ill.	29,259	41,024	56,100	66,950	129
St. Louis, Mo.	350,518	451,770	575,238	687,029	96
St. Paul, Minn.	41,473	133,156	163,065	214,744	418
Tulsa, Okla.	—	—	1,390	18,182	—
Wichita, Kans.	4,911	23,853	24,671	52,450	968

23. For more on flash flood mortality during the 1970s, see Jean French et al.,
    "Mortality from Flash Floods: A Review of National Weather Service Reports,
    1969–81," *Public Health Reports* 98 (1983): 584–588.

24. There were 296 tornado-related deaths in mobile homes from 1981 through 1997,
    accounting for 35 percent of the total number of tornado deaths. Before 1985, the
    federal government did not keep track of deaths in mobile homes. I found figures
    on mobile home deaths during tornadoes between 1981 and 1984 in the following
    sources. Frederick P. Ostby and Larry F. Wilson, "The 1981 Tornado Season:
    Frequent Storms but Few Deaths," *Weatherwise* 35 (1982): 30; Edward W.
    Ferguson et al., "The Year of the Tornado," ibid., 36 (1983): 22; Edward W.
    Ferguson, Frederick P. Ostby, and Preston W. Leftwich, Jr., "Tornado: Few
    Deaths, Little Damage," ibid., 37 (1984): 28; Ferguson, Ostby, and Leftwich,
    "Tornadoes Cause Record Number of Fatalities," ibid., 38 (1985): 23. For the
    period from 1985 through 1987 I relied on Richard A. Wood, "Summary of 1985
    Natural Hazard Deaths"; Wood, "Summary of 1986 Natural Hazard Deaths"; and
    Brian E. Peters, "Natural Hazard Deaths in 1987 in the United States." For the
    period from 1988 through 1995, I relied on summaries of natural hazard deaths in
    the United States available from the NWS, Silver Spring, Md. The figures for
    1996 and 1997 are available at www.nws.noaa.gov/om/severe_weather.
        It is important to note that scientific evidence on the poor survival rate of those
    living in mobile homes struck by tornadoes was available as early as 1980. A study
    done of a tornado that struck Wichita Falls, Texas, in 1979 demonstrated this and
    called on the government to amend current safety regulations "to decrease the impact
    of future tornadoes on human health." See Roger I. Glass et al., "Injuries from the
    Wichita Falls Tornado: Implications for Prevention," *Science*, Feb. 15, 1980, 735.

25. "Major Tokyo Quake Would Cost $1.2 Trillion, Study Says," *NYT*, Sept. 20, 1994.

26. "The City Beneath the Sea: New Orleans," *Weatherwise* 48 (1995): 56.

27. "A Sense of the Unreal in New Orleans Flood," *NYT*, May 14, 1995. Hurricane
    Betsy in 1965—only a category 3 storm—generated a 12-foot wave that came bar-
    reling out of Lake Pontchartrain and snapped telephone poles, capsized freighters,
    and killed more than 50 people.
        In considering the relative threat of hurricanes to human lives, it is impor-
    tant to note how disjointed the federal policy response has been in this regard,
    quite in contrast to the efforts made to mitigate the effects of earthquakes. Since
    1977, the federal government has provided money and resources to create the
    National Earthquake Hazards Reduction Program, an initiative that has sup-
    ported scientific research to improve the prediction and mitigation of seismic risk.
    However, no such national program for hurricane reduction has been created,
    save for a very recent and halfhearted attempt on FEMA's part. According to
    Thomas Birkland, the future prospects for a national hurricane policy initiative
    are very dim. In part, this is because of the distribution of the two different
    threats. Hurricanes affect large numbers of states, making it likely that all kinds of
    political opposition will emerge to mitigation, whereas the risk of earthquake, he
    explains, is more localized, making it likely that the government will be able to

design a mitigation program by working in a one-on-one relationship with, say, a state like California. In the end, Birkland concludes, "hurricanes currently are considered a land-use problem about which the federal government can do relatively little, while earthquakes are considered a scientific problem about which the federal government can do considerably more." Thomas A. Birkland, "Factors Inhibiting a National Hurricane Policy," *Coastal Management* 25 (1997): 399–400.

FOUR · BUILDING FOR APOCALYPSE

1. "Building Code Is Blamed for Keys Tragedy," *MH*, Sept. 13, 1960.
2. Stetson said the statement applied to 75 percent of the state of Florida. John Stetson, "Message from the President," *Florida Architect*, Feb. 1960, 17.
3. See table 4 in Fred Doehring, Iver W. Duedall, and John M. Williams, *Florida Hurricanes and Tropical Storms: 1871–1993: An Historical Survey*, Technical Paper no. 71 (Gainesville: Florida Sea Grant College Program, 1994), 61–67. Fifty-one people died in the 1947 hurricane. The table also notes that 75 were killed in Hurricane Betsy, although only 4 of those deaths occurred in Florida; the bulk of the deaths happened in Louisiana. Agnes in 1972 killed 122, but again, only 6 people in Florida lost their lives.
4. Figures are derived from table 2 in ibid., 53.
5. See, e.g., "Hurricane Expert Sees Calamitous Storms in U.S. Future," *Chronicle of Higher Education*, Dec. 14, 1994.
6. See table 4 in Doehring, Duedall, and Williams, *Florida Hurricanes*, 60, 63–65.
7. The following description of development in Miami Beach is based on Polly Redford, *Billion-Dollar Sandbar: A Biography of Miami Beach* (New York: Dutton, 1970), 204, 216, 255; Ruby Leach Carson, "Forty Years of Miami Beach," *Tequesta* 15 (1955): 23–25; and "Let's Go to the Beach—But Where Did It Go?" *MDN*, Sept. 21, 1958. The hotel boom described here was followed by a surge in high-rise apartments in the 1960s as the Seacoast Towers, Manhattan Towers, Regency Tower, and Surfside Tower, among others, filled in the waterfront even more.
8. Melvin J. Richard, "Tidelands and Riparian Rights in Florida," *Miami Law Quarterly* 3 (1949): 349, 359.
9. *State* v. *Simberg*, 2 Fla. Supp. 178 (Dade Cty. Cir. Ct., 1952); *State* v. *Simberg*, 4 Fla. Supp. 85 (Dade Cty. Cir. Ct., 1953).
10. "Public Won't Regain Use of Beaches Despite Court Ruling Against Hotels," *MDN*, Aug. 16, 1953.
11. Luther J. Carter, *The Florida Experience: Land and Water Policy in a Growth State* (Baltimore: Johns Hopkins Univ. Press, 1974), 144, 151–152; Edward Sofen, *The Miami Metropolitan Experiment* (Bloomington: Indiana Univ. Press, 1963), 15–16.
12. See Theodore E. Haeussner, "Tides and Flooding Incident to Winds of Hurricane Force," in Florida Hurricane Damage Study Committee, *Florida Hurricane Survey Report 1965* (n.p., n.d.), 67.
13. See *Atlantis Development Corp.* v. *United States*, 379 F.2d 818, 820–821 (5th Cir. 1967); and *United States* v. *Ray*, 294 F. Supp. 532, 535 (S. D. Fla. 1969).

14. Quoted in 294 F. Supp. at 535.
15. Quoted in "A Storm Brews in the Barge's Wake," *MH*, Mar. 14, 1965.
16. "Reef Cities Destroyed by Betsy," *MH*, Sept. 16, 1965.
17. 294 F. Supp. at 538; *United States* v. *Ray*, 423 F.2d 16 (5th Cir. 1970).
18. The following description of development in the keys is based on Charlton W. Tebeau, *A History of Florida* (Coral Gables: Univ. of Miami Press, 1971), 403–404; and Oliver Griswold, *The Florida Keys and the Coral Reef* (Miami: Graywood Press, 1965), 70–72.
19. Peter J. May, *Recovering from Catastrophes: Federal Disaster Relief Policy and Politics* (Westport, Conn.: Greenwood, 1985), 23, 27.
20. Griswold, *Florida Keys*, 121.
21. Quoted in "Betsy Tops Cleo in Home Damage," *MH*, Sept. 15, 1965.
22. The hurricane also led to legislation requiring HUD to study the nation's flood problem, which it did, ultimately recommending passage of a federal flood insurance program. See tables 2.2 and 2.3 in May, *Recovering from Catastrophes*, 28, 33.
23. "Disaster? Not in Miami," *MH*, Aug. 29, 1964.
24. "A 'Disaster' to Think About," *MDN*, Aug. 31, 1964.
25. "Beaches Are Washing Away, Golden Foot by Foot," *MDN*, Aug. 9, 1959.
26. Orrin H. Pilkey Jr. et al., *Living with the East Florida Shore* (Durham, N.C.: Duke Univ. Press, 1984), 44.
27. "Public Beach Access 'Eroded' Away," *MH*, Oct. 19, 1967.
28. "Beaches Are Holes Near the Water into Which One Pours Money," *MH Tropic Magazine*, July 10, 1977, 18.
29. Ibid., 23.
30. Quoted in "The Sands of Time Are Running Out for Miami Beach," *NYT*, May 10, 1970.
31. Quoted in "Judge Questions Miami Beach's Right in Erosion Suit," *MH*, Dec. 21, 1969.
32. "Hotels Offer Own Plan to Save Beachfronts," *MH*, Oct. 21, 1967.
33. "A Mayor Covers the Waterfront," *MDN*, Nov. 10, 1967.
34. "No Beach Is Good Beach?" *MH*, Dec. 4, 1965.
35. Quoted in "A Mayor Covers the Waterfront."
36. Ben Novack, "'It's No Place for Picnics,'" *MH*, May 29, 1970.
37. "Hotels OK Plan for 10-Mile Public Beach," *MH*, Oct. 14, 1971.
38. Pilkey et al., *East Florida Shore*, 37.
39. "Believe It: S. Florida Short on Sand," *MH*, Oct. 25, 1993.
40. Hurricanes Donna (category 4) and Betsy (category 3), though considered major storms, did not directly strike Miami. Hurricane Cleo, which did directly hit the area, was only a category 2 storm.
41. Prior to the Second World War, the municipalities of Miami, Miami Beach, and Coral Gables each had their own separate building codes. But in 1957, the area's building interests put pressure on public officials to rationalize this complex set of codes in an effort to save them money. Sofen, *Miami Metropolitan Experiment*, 20, 26.

42. Shortly after it became the law, Gordon Dunn, the chief forecaster in the Weather Bureau's Miami office, called the code, "the best hurricane building code in the world." Quoted in Ivan H. Smith, "Building Codes and Hurricane Damage," in *Hurricane Survey Report 1965*, 49. Unfortunately, however, the code, whatever its merits, was laxly enforced. This became especially evident after hurricanes Donna and Betsy, when studies commissioned by the state showed evidence of complacency among builders. The studies, however, were roundly ignored in the spirit of South Florida hurricane denial. See *Florida Hurricane Report Concerning Hurricane "Donna"* (n.p., 1961); *Hurricane Survey Report 1965;* and "Donna's Lessons from 1960 Ignored," *MH*, Sept. 20, 1992.

43. After Hurricane Andrew, the *Miami Herald* reviewed the minutes of the Board of Rules and Appeal, which interprets the building code for Dade County. The *Herald* found that the board, dominated by building interests, had "repeatedly given in to a construction industry looking for cheaper, quicker ways to build." In 1968, the board even went so far as to recommend that the county sue Coral Gables because the municipality, concerned to maintain the highest building standards, refused to abide by the less rigorous criteria of the South Florida Building Code. In 1972, the idea of including workmanship standards in the code was first broached. But the board spent 20 years dragging its feet before deciding to include some quality assurance. See "Building Code Eroded over Years," *MH*, Oct. 11, 1992.

44. Quoted in "Builders' Shortcut to Disaster," *MH*, Dec. 20, 1992.

45. Alan Sirkin, "Engineering Overview of Hurricane Andrew in South Florida," *Journal of Urban Planning and Development* 121 (1995): 6.

46. Quoted in "Andrew Exposes Safety Gaps," *Engineering News-Record*, Sept. 7, 1992, 11.

47. In Dade and Broward counties, the median household income of those living in mobile homes is one-third less than that of households occupying conventionally built homes. See "Despite Storm's Carnage, Mobile Home Sales Are Brisk," *MH*, Sept. 26, 1992.

48. *Report Concerning Hurricane "Donna"*, 18.

49. "It has been very evident," wrote architect John Stetson after the storm, "that winds of 60 m.p.h. or greater are destructive to trailer parks." *Hurricane Survey Report 1965*, 53, 74.

50. Louis Frey and J. Richard Knop, "The Imperative Need for Uniform Mobile Home Safety Standards," *Washington and Lee Law Review* 30 (1973): 462n. The authors note that eight mobile home residents were killed during Agnes. The figure on the total number of Floridians killed in the disaster was derived from Jay Barnes, *Florida's Hurricane History* (Chapel Hill: Univ. of North Carolina Press, 1998), 239.

51. For more on the way that industry dominated the American National Standards Institute, see Senate Committee on Banking, Housing, and Urban Affairs, *1973 Housing and Development Legislation*, 93d Cong., 1st sess., 1973, Committee Print, 880, 1137, 1218–1219.

52. Frey and Knop, "Mobile Home Safety," 473n.

53. For a brief history of manufactured housing regulation, see National Commission on Manufactured Housing, *Final Report* (Washington, D.C.: GPO, 1994), 19–20.

For more on the politics of preemption, see Senate Committee on Banking, Housing, and Urban Affairs, *1973 Housing and Development Legislation*, 862–891. To this day, preemption remains the industry's sacred cash cow. It allows manufacturers to produce homes with a minimum of regulatory interference, quite in contrast to the myriad of state and local building ordinances covering conventionally built houses. Under the federal code, for example, manufacturers are not required to have a licensed plumber or electrician on-site. When state and local governments have tried to impose this entirely reasonable requirement, the industry has successfully asserted preemption. It's no wonder that in announcing its goals for 1998, the Manufactured Housing Institute, the industry's most powerful trade group, put maintaining preemption at the top of the list.

54. See James R. McDonald and John F. Mehnert, "Review of Standard Practice for Wind-Resistant Manufactured Housing," *Journal of Aerospace Engineering* 2 (1989): 90–91.

55. See the testimony of Peter R. Sparks in House Committee on Banking, Finance, and Urban Affairs, *Manufactured Housing*, 102d Cong., 2d sess., 1992, Committee Print, 22.

56. "Safety Claims for Mobile Homes Called a Dangerous Myth," *MH*, Sept. 7, 1992.

57. As late as 1992, the FMHA's reference guide noted, "The wind load factor of HUD Code homes also often exceeds local building code requirements. The strict HUD Code standards have resulted in manufactured homes withstanding 110-mph hurricane winds. Manufactured homes built to the HUD Code and anchored properly are as safe as any other structure in high winds." Quoted in "Andrew's Ruins Say It All About Mobile Homes," *St. Petersburg (Fla.) Times*, Oct. 5, 1992.

58. McDonald and Mehnert, "Manufactured Housing," 91.

59. Quoted in "Fed's Failed to Heed Mobile Home Warnings," *MH*, Sept. 21, 1992.

60. The percentage figure, provided by HUD, is from the facts in *Florida Manufactured Housing Ass'n, Inc.* v. *Cisneros*, 53 F.3d 1565, 1569 (11th Cir. 1995). Similarly, Hurricane Hugo destroyed or damaged 90 percent of the manufactured housing in the 24-county area of South Carolina hit hardest by the storm. See Mary C. Comerio, *Disaster Hits Home: New Policy for Urban Housing Recovery* (Berkeley: Univ. of California Press, 1998), 51, 55.

61. These figures are from "Safety Claims for Mobile Homes."

62. House Committee on Banking, Finance, and Urban Affairs, *Manufactured Housing*, 35.

63. The figure of 130 mph is a "fastest-mile" wind speed, that is, the time it takes wind at 10 meters above the earth's surface to journey one mile. See "Clocking Andrew's Awful Gusts," *Engineering News-Record*, Nov. 23, 1992, 8. Roger Pielke Jr., an authority on the hurricane, has written: "Beyond the relatively small area of greater than 120 mph winds, most of Dade County experienced sustained wind speeds of less than 120 mph....Thus, Andrew was not a force of nature beyond experience." A small area near the storm's center, Pielke explains, had winds of 145 mph with gusts of 175. Roger A. Pielke Jr., *Hurricane Andrew in South Florida:*

Mesoscale Weather and Societal Responses (Boulder, Colo.: National Center for
Atmospheric Research, 1995), 137.

64. For an excellent discussion of the debate over how to measure Andrew's wind
    speed, see "Clocking Andrew's Awful Gusts."

65. "U.S. Experts Fault Code on Mobile Home Safety," *MH*, Sept. 10, 1992.

66. It also bears pointing out that some, including Richard Wright of the National
    Institute of Standards and Technology, see the improved HUD wind standard as
    simply an "interim measure." "Important questions regarding the behavior of
    building envelopes under intense fluctuating pressures, their susceptibility to pen-
    etration by wind-generated missiles, and the consequences of increased internal
    pressures will remain." See his testimony in House Committee on Science, Space,
    and Technology, *Affordable Housing and Construction R&D*, 103rd Cong., 1st sess.,
    1993, Committee Print, 10.

67. "New Standards Worry Mobile Home Makers," *Fresno (Calif.) Bee*, Jan. 23, 1994.

68. Quoted material is from 53 F.3d at 1565, 1582–1583.

69. Quoted in "A Sturdier Manufactured Home," *Orlando (Fla.) Sentinel*, Oct. 16,
    1994.

70. "Nobility Homes Keeps Market in Its Mind," *Tampa (Fla.) Tribune*, May 12,
    1996.

71. Quoted in "When a Home Isn't a House," *St. Petersburg Times*, Feb. 16, 1992.

72. Quoted in "Mobile Homes Are Still Battling Tin-Can Image," ibid., Sept. 13,
    1988.

73. "Manufactured Homes," *Orlando Sentinel*, Letter to the Editor, Nov. 7, 1994.

74. Quoted in "For Safer Mobile Homes," *St. Petersburg Times*, Mar. 19, 1994.

75. Quoted in "In Mobile Homes, Sitting Vulnerable to Tornadoes," *NYT*, Mar. 2,
    1998.

76. Quoted in the written testimony before Congress of James Phillips, director of
    the Manufactured Housing Division of the Missouri Public Service Commission,
    in House Committee on Banking, Finance, and Urban Affairs, *Manufactured
    Housing*, 127. According to Phillips, a mobile home manufacturer told this to a
    Missouri state official.

    Although official figures from the NWS are not in as of this writing, it
appears that 1998 was the worst year for tornado-related fatalities in nearly two
decades. Yet that fact did not stop the House of Representatives from passing in
the fall of that year a thoroughly proindustry, antisafety piece of legislation in the
name of modernizing the manufactured housing act of 1974. The legislation,
which failed to gain passage in the Senate, threatened to gut what little protection
the 1974 law affords by yoking regulatory decisions to an economic calculus. The
bill eliminated the mandate requiring the federal government to regulate these
structures with an eye toward reducing "the number of personal injuries and
deaths." In its place, the modernized legislation would have forced the government to promote the acceptance and affordability of manufactured housing, com-
promising safety in the process. HUD was able to prevail in court on the new
wind standards in part because the 1974 act gives the agency broad responsibility

for protecting not just mobile home residents but others who, say, might be injured by flying debris. The bill passed by the House, however, only required that HUD protect mobile home owners. It also reduced the standard of protection from "highest" to simply "high." Finally, it required that preemption be broadly construed, a move that, according to HUD, could force it to rescind local safety regulations like those requiring fire sprinklers.

FIVE · UNCLE SAM–FLOODPLAIN RECIDIVIST

1. Bodley added: "This one [the flood] was beyond the control of anybody." Quoted in "In the Path of the Flood, Hard Times Just Got Harder," *NYT*, Aug. 8, 1993.
2. "Troubled Waters," *Newsweek*, July 26, 1993, 20.
3. Lee Wilkins, "Living with the Flood: Human and Governmental Responses to Real and Symbolic Risk," in *The Great Flood of 1993: Causes, Impacts, and Responses*, ed. Stanley A. Changnon (Boulder, Colo.: Westview Press, 1996), 224.
4. *History of St. Charles, Montgomery, and Warren Counties, Missouri* ....(1885; reprint, St. Louis: National Historical Co., 1969), 136.
5. Walter B. Stevens, *Centennial History of Missouri: One Hundred Years in the Union 1820–1921* (St. Louis: S.J. Clark Publishing Co., 1921), 242.
6. "The Great Flood," *St. Charles Cosmos*, May 25, 1892.
7. "Birdseye View of the Vast Flooded District," *SLPD*, June 7, 1903.
8. Quoted in Lori Breslow, ed., *Small Town* (St. Charles: John J. Buse Historical Museum, n.d.), 60.
9. A useful list of historic floods in St. Charles can be found in "Record of Past Floods Listed," *St. Charles (Mo.) Banner-News*, Mar. 30, 1973. The list, however, fails to mention the 1915 and 1917 floods.
10. Fredrick J. Dobney, *River Engineers on the Middle Mississippi: A History of the St. Louis District, U.S. Army Corps of Engineers* (Washington, D.C.: GPO, 1978), 92–96.
11. See "'Slumlord' of the Flood Plain," *SLPD*, Feb. 20, 1994.
12. Darren L. May, associate planner, St. Charles County Department of Planning, telephone interview with author, Dec. 17, 1996.
13. Quoted in "Birth of Interstate to Be Marked Saturday," *St. Charles (Mo.) Post*, Aug. 9, 1996.
14. "Zoning Vital to Orderly Growth of Co. Population," *SCJ*, Dec. 18, 1958.
15. Telephone interview with Darren May.
16. The Missouri River gauge at St. Charles is located 27.8 river miles upstream from the confluence of the Mississippi River. I was able to derive these figures on the number of times the gauge exceeded 30 feet from data graciously compiled for me by Jack Burns, senior service hydrologist, NWS Forecast Office, St. Louis.
17. Quotations are from "Levee Plan Is Attacked," *SCJ*, Oct. 4, 1971.
18. "Floods of Controversy Pour over Proposal to Build Levee," *St. Louis Globe-Democrat*, Sept. 27, 1971; "Levee Hearing Here Tonight," *SCJ*, Sept. 30, 1971; "Levee Opponents, Backers Clash at Hearing," *St. Charles Banner-News*, Oct. 1, 1971.

19. "Floods Revive Dispute over Levee Plan," *St. Louis Globe-Democrat*, June 23, 1973.
20. Quoted in "Levees—Has the Flooding Taught Us a Lesson?" ibid., June 25, 1973.
21. See "Table Compares 1973, Earlier Major Floods," ibid, June 25, 1973.
22. For a brief discussion of the development of the National Flood Insurance Program, see Peter J. May, *Recovering from Catastrophes: Federal Disaster Relief Policy and Politics* (Westport, Conn.: Greenwood, 1985), 34.
23. "Flood Insurance Popularity Skyrockets," *St. Charles Banner-News*, Feb. 25, 1975.
24. "Court Meets Flood Ins. Deadline," *SCJ*, Dec. 18, 1978.
25. "100-Year Flood Standard Misstates Risk," *St. Louis Business Journal*, June 19, 1995. A building in a 100-year floodplain has, during the period of a 30-year mortgage, a 26 percent chance of sustaining damage from a 100-year or larger flood event, versus a 17 percent chance of incurring damage from a fire.
26. The town joined a number of other river communities in a lawsuit that claimed that the National Flood Insurance Program violated the Fifth Amendment by depriving it of property without due process. "Flood Insurance Suit Filed," *St. Charles Banner-News*, Nov. 15, 1977.
27. Quoted in "Federal Agency to Check Variances in Flood Plain," *St. Charles Post*, Mar. 3, 1982.
28. The crest in 1951 was 37.3 feet.
29. State Sen. Steve E. Ehlmann of Missouri, telephone interview with author, Dec. 12, 1996.
30. Quoted in "Missouri Could Rise Again, Officials Warn," *St. Charles Banner-News*, Mar. 29, 1978.
31. Quoted in "Floods Are Ritual for County Areas," *St. Charles Post*, Mar. 26, 1982.
32. Figures are from "Flood Insurance Jeopardized, County Warned," ibid., Apr. 6, 1982.
33. "Interagency Hazard Mitigation Report for Missouri" (Prepared by FEMA in Nov. 1986 in response to the Oct. 14, 1986, disaster declaration), 4. The crest in 1986 was two-tenths of a foot higher than the 1951 crest.
34. Figures are from "Receding Flood Bares Debate over Mobile Home Parks," *SLPD*, Nov. 27, 1986.
35. "Interagency Hazard Mitigation," 2. Only two other recorded instances of such a detour exist: once in 1951 and again in 1973.
36. After the 1986 flood, FEMA and the corps sat down at the drawing board to sketch a new map of St. Charles County. This time they drew a 12-mile long "natural overflow channel" between the Mississippi and Missouri rivers. That was the path that some of the water had taken in 1986 (and in 1951 and 1973). Their motive in drawing the map was to establish new floodplain regulations relating to the channel that would further dim the prospects for the farmers' dream levee. In the end, the overflow channel idea turned out to be more of a threat than something FEMA and the corps were dead set on. But the farmers needed to take it seriously all the same. So they set out to show that there was really nothing natural about the channel or the 1986 flood. What they had on their hands here, they said, was a man-made disaster—a company-made one to be more precise. The company at issue was the Union Electric Company, operators of the Bagnall Dam on the Osage River. Apparently

the company had not planned for the wet weather preceding the 1986 flood, and as a result it needed to open the floodgate on its dam to prevent overflowing its property. The additional water released by the dam then coursed downstream, causing a second flood crest as the Osage's water poured into the Missouri and forged downstream to St. Charles, wrecking some 300 agricultural levees on the way. In its defense, the company explained that its dam produced electricity and was not by any stretch of the imagination a flood control project. "If they want to blame anyone for this," said company spokesman Thomas Dehner, referring to the farmers, "blame Mother Nature." Quoted in "Army Engineers Should Broaden Use of Bagnall Dam, Farmers Say," *SLPD*, May 21, 1990.

Obviously, the 1986 flood was not simply the result of natural forces. Nothing could possibly be so simple and clear-cut in a watershed where tremendous development and flood control had occurred for some 200 years. And, of course, to blame simply Union Electric for the destruction, as the farmers did, was also shortsighted, not to mention self-serving. St. Charles farmers were quick to point out the unnaturalness of the disaster only to the extent that it furthered their own unnatural ambitions for a levee—one that would certainly lead to more development and more destructive floods in the future.

37. See FEMA, National Flood Insurance Program, Proposed Rules, *Federal Register*, 51, no. 60 (Mar. 28, 1986): 10742–10743.
38. Quoted in "Second Public Hearing Likely," *SCJ*, Jan. 11, 1987.
39. Quotations are from ibid.
40. Quoted in "Flood Plain Regulations Given Reluctant Recommendations," *SCJ*, Mar. 8, 1987.
41. "Congress May Delay New Flood Plain Rules," *SCJ*, Apr. 8, 1987.
42. "New Flood Rules Could Backfire, Engineer Says," *SLPD*, May 1, 1990.
43. Quotations are from "State Steps in to Check Flooding Problem," *SLPD*, Feb. 8, 1988.
44. "Midwest Flooding: Communities with 10 or More Repetitive Loss Structures in Declared Counties" (Compiled by FEMA, Region 7, Kansas City, Mo., Aug. 27, 1993).
45. Quoted in "Going Back for More When It Floods in St. Charles County," *SLPD*, Nov. 21, 1993.
46. Quoted in "On the Disaster Dole," *Newsweek*, Aug. 2, 1993, 24.
47. Tom Szilasi, building commissioner (1989-1994), St. Charles County, telephone interview with author, Dec. 4, 1996.
48. Quoted in "2 Legislators Decry U.S. Agency's Plans on Flood Insurance," *SLPD*, Jan. 21, 1991.
49. Telephone interview with Steve Ehlmann.
50. To help resolve the impasse over this issue, FEMA came up with the "density floodway" concept. Technically speaking, the floodplain was divided into two parts: a floodway fringe, where new construction would have to meet the one foot over the 100-year-flood elevation standard, and the floodway proper, where no future building at all could take place. The idea behind the density concept was

to divide up the bona fide floodway into one-half-mile squares and allow development to fill 18 percent of each block, as long as all new structures were built one foot above the 100-year flood elevation. The catch was that the density floodway extended over a much larger area of the county. So the choice was simple: have a regular floodway that would cover less area but allow no new building at all, or a density floodway that would extend over more land and allow 18 percent development. Farmers would lose either 100 percent of their power to build or 82 percent. The county chose the density floodway plan late in 1992, though it was not a move sanctioned by local State Rep. Joseph Ortwerth, a long-time foe of floodplain management. Ortwerth explained that agreeing to the density floodway designation would turn the triangle "into a backwater basin, a cesspool for flood water. They are asking county government to sign a death warrant for this part of the county. It's a man-made disaster you're doing." Quoted in "Insurance Maps for Flood Plain Assailed, Ok'd," *SLPD*, Dec. 15, 1992.

51. See Miriam Anderson, "Midwest Floods of 1993: St. Charles Experience," *Forum for Applied Research and Public Policy* 11 (1996): 128.

52. The 500-year-flood level is 40.5 feet. On Aug. 2, 1993, the Missouri at St. Charles crested at 40.04 feet.

53. Quoted in "Agency's Tenants Can't Collect—But They Do," *SLPD*, Nov. 21, 1993.

54. Ron McCabe, branch chief, FEMA, Region 7, Kansas City, Mo., telephone interview with author, Nov. 27, 1996.

55. Quoted in "Agency's Tenants Can't Collect." The corps actually banned its tenants from collecting flood insurance. But the Federal Insurance Administration, which administers the flood insurance program, renewed the policies for fear of engaging in racial discrimination. After the flood, FEMA and the corps agreed that the county had the responsibility of doing damage assessments for the clubhouses. To date more than half the cabins have been torn down. Steven G. Lauer, director of planning, St. Charles County Department of Planning, telephone interview with author, Nov. 22, 1996.

56. Miriam G. Anderson, associate planner, St. Charles County Department of Planning, telephone interview with author, Dec. 12, 1996.

57. Rutherford H. Platt and Claire B. Rubin, "Stemming the Losses: The Quest for Hazard Mitigation," in Rutherford H. Platt, *Disasters and Democracy: The Politics of Extreme Natural Events* (Washington, D.C.: Island Press, 1999), 93–96.

58. Figure is from "St. Charles Moods Cover Gamut from Buoyant to Sinking," *SLPD*, July 11, 1993.

59. Quoted in "Mobile 'Homeless' Residents Wary of Official Plans," *SLPD*, Oct. 26, 1993.

60. Telephone interview with Steve Ehlmann.

61. The St. Louis County levees at Riverport, Earth City, and Chesterfield exist outside the regulatory floodway. The corps, despite some initial reservations, approved permits for them all. For more on the Riverport project and its implications for flooding in St. Charles, see "Riverport Project Opponents Cite Flood Plain Dangers," *SLPD*, Dec. 22, 1983; "Corps Planning Hearing on Flood Plain

Project," *SLPD*, Aug. 27, 1984; "'Riverport' Proposal Opposed at Hearing," *SLPD*, Oct. 2, 1984; and "Army Corps of Engineers Issues Permit for Riverport Project," *SLPD*, May 2, 1985.

62. Quoted in "Flood Planning Awaits End of 'Great' One," *SLPD*, Nov. 9, 1993.

63. Quotations are from "Officials Seek Relief from FEMA Rules," *SLPD*, Aug. 17, 1993.

64. Telephone interview with Darren May.

65. Telephone interview with Steve Ehlmann.

66. Shelia Harris-Wheeler, chair of the Unmet Needs Committee, telephone interview with author, Dec. 4, 1996.

67. Miriam Mahan, director of St. Joachim and Ann Care Services, telephone interview with author, Dec. 4, 1996.

68. See, e.g., Missouri State Emergency Management Agency, "Out of Harm's Way: The Missouri Buyout Program" (n.p., n.d.).

69. Quotations are from "Flood Aid Missing, Some Say Mobile-Home Parks Are Getting Left Out," *SLPD*, July 5, 1994.

70. Quoted in "Open with Closings First Buyouts in Flood Plain Begin, Officials Told," *SLPD*, May 13, 1994.

71. See "Flood Plain Plans Anger City Officials," *SLPD*, Dec. 16, 1993.

72. See "2 from Council Fight Proposal to Build Homes," *SLPD*, Jan. 31, 1994; and "Opposition Builds to Housing Plan," *SLPD*, May 13, 1994.

73. Miriam Gradie Anderson and Rutherford H. Platt, "St. Charles County, Missouri: Federal Dollars and the 1993 Midwest Flood," in Platt, *Disasters and Democracy*, 237.

74. See "Congress Raises Levee Protection," *SLPD*, Oct. 8, 1996; and the actual legislation, the Water Resources Development Act of 1996, 104 P.L. 303, § 547.

INTERLUDE: THE PERILS OF PRIVATE PROPERTY

1. This point is made in Douglas C. Dacy and Howard Kunreuther, *The Economics of Natural Disasters: Implications for Federal Policy* (New York: Free Press, 1969). Although figures on property damage are often unreliable, the authors note that there is little question that natural disasters caused increasing losses in the last two-thirds of the twentieth century. Average annual damages from hurricanes, floods, tornadoes, and earthquakes in the decades of the 1930s and 1940s were in the $350 million range (1964 dollars). In the 1950s, that figure climbed to over $600 million. See table 1.1 in ibid., 6. A more recent survey of just flood-related losses uncovered average annual damages from the 1920s through the 1940s of just under $800 million (1985 dollars). In the 1950s that figure increased to over $1.8 billion, retreating somewhat to $1.2 billion the following decade, and rising again to $3.9 billion in the 1970s. See table 3.8, "National Weather Service Estimates of Flood-Related Damages in the United States, 1916–1989," in *Floodplain Management in the United States: An Assessment Report*, prepared for the Federal Interagency Floodplain Management Task Force by L.R. Johnston Associates (1992), 2: 3–18.

2. This information is as of Dec. 1999 and was provided by the Insurance Information Institute, New York. The 10 most costly insured catastrophes are Hurricane Andrew, Aug. 1992 ($15.5 billion); the Northridge earthquake, Jan. 1994 ($12.5 billion); Hurricane Hugo, Sept. 1989 ($4.2 billion); Hurricane Georges, Sept. 1998 ($2.9 billion); Hurricane Opal, Oct. 1995 ($2.1 billion); the 20-state winter storm, Mar. 1993 ($1.8 billion); the Oakland, California fire, Oct. 1991 ($1.7 billion); Hurricane Fran, Sept. 1996 ($1.6 billion); Hurricane Iniki, Sept. 1992 ($1.6 billion); tornadoes, wind, and hail in Oklahoma, Kansas, and 16 other states, May 1999 ($1.5 billion). It is important to note, however, that the increase in damage was not a constant trend upward. Roger Pielke and Christopher Landsea have demonstrated that hurricane damage during the 1970s and 1980s was actually less than in earlier decades. Only in the 1990s do the high damage levels approach the level of the 1940s through the 1960s. Roger A. Pielke Jr. and Christopher W. Landsea, "Normalized Hurricane Damages in the United States: 1925–1995," (Draft paper, June 10, 1997, available at: www.dir.ucar.edu/esig/HP_roger/hurr_norm.html).

3. For more on the Decade for Natural Disaster Reduction, see National Research Council, *A Safer Future: Reducing the Impacts of Natural Disasters* (Washington, D.C.: National Academy Press, 1991); Gilbert F. White, "A Perspective on Reducing Losses from Natural Hazards," *BAMS* 75 (1994): 1237–1240; G.O.P. Obasi, "WMO's Role in the International Decade for Natural Disaster Reduction," *BAMS* 75 (1994): 1655–1661; and James P. Bruce, "Natural Disaster Reduction and Global Change," *BAMS* 75 (1994): 1831–1835.

4. Quoted in "The End of an Era in Homeowners," *Best's Review* (Property/Casualty Edition), May 1994, 22.

5. "Redlining the Coasts," *Audubon*, July/Aug. 1993, 16.

6. The next longest lull in severe storm activity lasted 12 years and occurred between 1935 and 1947. See table 2 in *The Impact of Catastrophes on Property Insurance* (New York: Insurance Services Office, 1994), 7.

7. Consider, for example, that the combined population in 1990 of Dade, Broward, and Palm Beach counties was more than four million. That was a higher total than that of 29 states. The trend in property growth, however, is perhaps even more stunning. As of 1993, insured property in 168 coastal counties from Texas to Maine totaled over $3.1 trillion, a 50 percent increase (adjusted for inflation) over the 1988 total. See Roger A. Pielke Jr., *Hurricane Andrew in South Florida: Mesoscale Weather and Societal Responses* (Boulder, Colo.: National Center for Atmospheric Research, 1995), 56, 64.

8. Quoted in ibid., 6. "The insurance system is just not built to handle the once-in-a-100-year event with $50 billion or $100 billion of losses," remarked Sean Mooney, chief economist at the Insurance Information Institute. Quoted in "Florida Insurers Still Face a Gap in Disaster Coverage," *NYT*, Aug. 8, 1995.

In an attempt to rescue the insurance industry from the threat of a major calamity, various insurance groups, in league with some consumer organizations, have formed the Natural Disaster Coalition. The coalition has been lobbying

Congress to set up a disaster trust fund that insurers could dip into for help during especially calamitous years. Such efforts to involve the federal government have had little success as yet, prompting others to hitch their hopes for a solution to the catastrophe problem to the free market. Rather than have the government underwrite the risk of disaster, one idea is to allow speculators to purchase that risk by investing in insurance derivatives. The Chicago Board of Trade has recently set up a calamity index that is based on the amount paid out in insurance claims divided by premiums. Brokers can write futures and options contracts against the index. Essentially, investors are rolling the dice on whether it's going to be a good or a bad year for natural disasters. Insurance derivatives are just another way of reshuffling the risk of calamity. Moreover, they are likely to promote the same kind of dangerous behavior that federal subsidies have encouraged. See "Deeper Pockets," *Forbes*, Sept. 26, 1994, 42–44; "New Tools Spread Risks of Insurers," *NYT*, May 15, 1995; and "Even Nature Can Be Turned into a Security," *NYT*, Aug. 6, 1997. For more on the Natural Disaster Coalition and its efforts in Congress, see "Is the Industry Courting Disaster?" *Best's Review* (Property/Casualty Edition), Mar. 1995; and "Pace Quickens in the Search for Help in Catastrophe Crisis," *Best's Review* (Property/Casualty Edition), July 1995.

9. "The Big Blow started as a little pout of hot air in the summer sky over West Africa," wrote *Newsweek* shortly after the disaster. A few paragraphs later, the magazine had puffed up nature's fury into what it labeled "a Category 5 catastrophe." In fact, Hurricane Andrew, although the most intense storm since Hurricane Camille in 1969, was just a category 4 storm. Seeking to further magnify the enormity of the event, the magazine likened the storm to that other great archetypal calamity of the twentieth century, the 1906 California earthquake. "The stories from Andrew's ground zero," the magazine explained, "summoned up a kind of terror that the country had not felt since the 1906 San Francisco earthquake." Moreover, in an effort to rationalize the federal government's poor response to what it described as overwhelming natural forces, *Newsweek* went on to conclude, "Andrew was what the insurance companies call 'an act of God,' a happening for which no mere human can be held to account; and Force 5 hurricanes are so rare, one or two every century or so, that it's no wonder people get out of practice at coping with them." After having described Andrew as beyond anyone's control, the magazine, in a fit of contradiction, ran a separate article in the very same issue on the role that human-induced global warming might play in transforming relatively benign storms into killer hurricanes. Although it did note that it was "impossible to blame any particular storm on global warming," one can only wonder what readers themselves made of the news magazine's tortured coverage. "What Went Wrong" and "Was Andrew a Freak—Or a Preview of Things to Come?" *Newsweek*, Sept. 7, 1992, 22, 23, 30.

A rather different optic on the "bigness" of the storm was offered by Kate Hale, Dade County's director of emergency management. "This was not the Big One of my nightmares," she said. The storm moved fast, carried relatively little water, and struck in an area where barely one-fifth of the population of the county lived. Quotation is from Walter Gillis Peacock, Betty Hearn Morrow, and Hugh Gladwin, *Hurricane Andrew: Ethnicity, Gender, and the Sociology of Disasters* (London: Routledge, 1997), 8.

10. "Contrary to many expectations that globally tropical cyclones may be becoming more frequent and/or more intense due to increasing concentrations of green-house gases, regionally the Atlantic basin has in recent decades seen a significant trend of fewer intense hurricanes and weaker cyclones overall." See Christopher W. Landsea et al., "Downward Trends in the Frequency of Intense Atlantic Hurricanes During the Past Five Decades," *Geophysical Research Letters* 23 (1996): 1700.

11. See Pielke and Landsea, "Normalized Hurricane Damages."

12. *Federal Disaster Assistance: Report of the Senate Task Force on Funding Disaster Relief* (March 15, 1995), 104th Cong., 1st sess., 1995, S. Doc. 104-4, 23.

13. Landsea et al., "Downward Trends," 1698.

14. *Floodplain Management in the United States: An Assessment Report*, prepared for the Federal Interagency Floodplain Management Task Force by the Natural Hazards Research and Applications Information Center (1992), 1: 24.

15. See "The High Risks of Denying Rivers Their Flood Plains," *NYT*, July 20, 1993.

16. See Executive Office of the President, *Sharing the Challenge: Floodplain Management into the 21st Century* (Report of the Interagency Floodplain Management Review Committee, Washington, D.C., 1994).

17. Quoted in "15 Floods in 25 Years, Town's Moving," *NYT*, June 2, 1994.

18. Quoted in "The High Risks of Denying Rivers Their Flood Plains."

19. Quoted in "After Flood, 2 Towns Diverge About the Next One," *NYT*, Aug. 7, 1993.

20. For information on the development of the island during the 1940s and 1950s, see "Beach Community Has Grown Rapidly," *Charleston Evening Post*, Jan. 13, 1953. For more on Long, see "J.C. Long: Couldn't Be Happier," *NC*, Jan. 27, 1980; and "Noted S.C. Developer, Landowner J.C. Long Dies," *NC*, July 10, 1984.

21. See, e.g., "Wise Critical of Dune Leveling," *NC*, Jan. 3, 1975; and "Dunes More Than Environmental Issue," *NC*, Jan. 20, 1975.

22. David Lucas, *Lucas vs. The Green Machine* (Alexander, N.C.: Alexander Books, 1995), 72.

23. On the politics behind the passage of South Carolina's beach protection act, see "Opposition to Beach Regulation Mounting," *State* (Columbia, S.C.), Jan. 17, 1988; "'Sellout' Beach Pact Protested," ibid., Mar. 8, 1988; "Out with the Tide: S.C. Beaches in Danger," ibid., May 22, 1988; and "Law to Stem S.C. Beach Erosion Called Too Little, Too Late by Some," *Atlanta Journal and Constitution*, June 5, 1988.

24. *Lucas v. South Carolina Coastal Council*, 304 S.C. 376, 404 S.E.2d 895 (1991); *rev'd*, 505 U.S. 1003 (1992).

25. Rutherford H. Platt, "Life After Lucas: The Supreme Court and the Downtrodden Coastal Developer," *Natural Hazards Observer* 17 (1992): 8; Gered Lennon et al., *Living with the South Carolina Coast* (Durham, N.C.: Duke Univ. Press, 1996), 183.

26. In addition, Lucas might possibly have damaged the sand dunes that protected property further inland.

27. 505 U.S. at 1072. For useful discussions of the case that explore its environmental consequences, see John Tibbetts, "Everybody's Taking the Fifth," *Planning*, Jan. 1995, 4–9; and Louise A. Halper, "A New View of Regulatory Takings?" *Environment*, Jan./Feb. 1994, 2–5, 39–40.

28. *Dolan* v. *City of Tigard*, 317 Ore. 110, 854 P.2d 437 (1993); *rev'd*, 512 U.S. 374, 405. For a discussion of the case, see Rutherford H. Platt, "Parsing *Dolan*," *Environment*, Oct. 1994.

29. "G.O.P. Pressing for New Property Rights Bill," *NYT*, Oct. 9, 1997. On the property rights movement more generally, see "Landowners Unite in Battle Against Regulators," *NYT*, Jan. 9, 1995.

SIX · THE NEUROTIC LIFE OF WEATHER CONTROL

1. Enact such a plan, Orville predicted, and within 40 years there would be "weather control to stagger the imagination." Quoted in "Our Mad Weather," *Newsweek*, Sept. 10, 1956, 62. Orville's address before the University Club of New York on Feb. 8, 1958, outlining his plan, is reprinted in *Congressional Record*, 85th Cong., 2d sess., 1958, 104, appendix: A1569–A1570. Under pressure from constituents who complained that weather modification companies were not living up to their promises, Congress passed legislation creating a presidential advisory committee on weather control. Appointed in 1953, the committee, headed by Orville, spent five years studying the efficacy of weather modification. It concluded that cloud seeding could be used effectively to increase precipitation by an average of 10 to 15 percent. That turned out to be an overly optimistic assessment, at least in the minds of some statisticians, who criticized the committee's handling of the evidence. See Robert G. Fleagle, ed., *Weather Modification: Science and Public Policy* (Seattle: Univ. of Washington, 1969), 10.

2. See "Weather Under Control," *Fortune*, Feb. 1948, 134; and National Research Council, *Weather and Climate Modification: Problems and Progress* (Washington, D.C.: National Academy of Sciences, 1973), 106.

3. See fig. 2 in Senate Committee on Commerce, Science, and Transportation, *Weather Modification: Programs, Problems, Policy, and Potential*, 95th Cong., 2d sess., 1978, Committee Print, 51.

4. "A Reporter at Large: Five-Ten on a Sticky Day," *New Yorker*, May 28, 1955, 40, 44, 70, 72; "Destroy Tornadoes by Guided Missile," *Science News Letter*, Nov. 21, 1953, 328; "No A-Bomb Weather Effect," *Science News Letter*, June 27, 1953, 397. Far more definitive proof that the atomic bomb did not cause the 1953 tornadoes is available in L. Machta and D.L. Harris, "Effects of Atomic Explosions on Weather," *Science*, Jan. 21, 1955, 78–79, 81. For more on the use of nuclear warheads to destroy tornadoes see Keay Davidson, *Twister: The Science of Tornadoes and the Making of an Adventure Movie* (New York: Pocket Books, 1996), 85–89.

5. Adm. Luis de Florez, "Weather—Take It or Make It" reprinted in *Congressional Record*, 87th Cong., 1st sess., 1961, 107, pt. 4: 5434.

6. Ray Jay Davis, "President's Message," *Journal of Weather Modification* 12 (1980): iv.

7. Edward A. Morris, "The Law and Weather Modification," *BAMS* 46 (1965): 621, 622.

8. Arthur L. Rangno and Peter V. Hobbs, "A New Look at the Israel Cloud Seeding Experiments," *Journal of Applied Meteorology* 34 (1995): 1169–1193. See also Richard A. Kerr, "Cloud Seeding: One Success in 35 Years," *Science*, Aug. 6, 1982, 519–521.

9. Vincent J. Schaefer, "A Call for Action," *Journal of Weather Modification* 2 (1970): 9.

10. Donald Worster, *Rivers of Empire: Water, Aridity, and the Growth of the American West* (New York: Pantheon, 1985), 243.

11. Robert L. Hendrick and Donald G. Friedman, "Potential Impacts of Storm Modification on the Insurance Industry," in *Human Dimensions of Weather Modification*, ed. W.R. Derrick Sewell (Chicago: Univ. of Chicago Press, 1966), 234, 236, 246. To their credit, the insurance men did question the desirability of weather control, noting that "storm benefits and losses cannot be completed by arithmetic."

12. Weather Modification Association, *Weather Modification: Some Facts About Seeding Clouds* (Fresno, Calif.: Weather Modification Association, 1977), 16.

13. The following account of the drought is based on "Drought: Growing U.S. Problem," *U.S. News & World Report*, Nov. 16, 1964, 92; "Parched," *Newsweek*, Nov. 4, 1963, 69; "A Fourth Dry Year for the Northeast?" *U.S. News & World Report*, June 7, 1965, 10; and "A Dry Silence in the Northeast," *Sports Illustrated*, Aug. 9, 1965, 22.

14. For a detailed discussion of Howell's cloud seeding and the legal case that ultimately grew out of the farmers' opposition to his efforts, see Theodore Steinberg, *Slide Mountain, Or The Folly of Owning Nature* (Berkeley: Univ. of California Press, 1995), 106–134.

15. Quoted in "Hail Control Meeting Attended by Farmers," *Mercersburg (Pa.) Journal*, June 22, 1962.

16. Hosler's version of these events is available in "Public Hearing on Weather Modification," Sept. 12, 1968, pp. 64–66, Records of the Pennsylvania Department of Agriculture, Bureau of Farmland Protection, Harrisburg, Pa. For more on Hosler's views about cloud seeding, see C.L. Hosler, "Overt Weather Modification," *Reviews of Geophysics and Space Physics* 12 (1974): 523–527.

17. Wallace E. Howell, "Cloud Seeding and the Law in the Blue Ridge Area," *BAMS* 46 (1965): 329.

18. "Weather—No Greater Responsibility," *Fulton Democrat* (Fulton County, Pa.), Sept. 10, 1964. Oakman was the editor of the *Fulton Democrat* and used the paper to promote his ideas about the wrongfulness of cloud seeding.

19. "Statement by the Penna. Natural Weather Association," *Mercersburg Journal*, Jan. 1, 1965.

20. "Weather—No Greater Responsibility."

21. "War to Win the Weather," Letter to the Editor, *Fulton Democrat*, Oct. 8, 1964.

22. "Co. Commissioner Hits Cloud Seeding," Letter to the Editor, *Fulton County (Pa.) News*, Sept. 24, 1964.

23. Trial Transcript, Cross-Examination of Delmar Mellott, p. 89, Court Papers filed in *Pennsylvania Natural Weather Ass'n* v. *Blue Ridge Weather Modification Ass'n,* Court of Common Pleas of Fulton County, Jan. Term, 1965, no. 3 in equity, Fulton County Courthouse, McConnellsburg, Pa.

24. "Big City Brother Feels Sorry for Poor Country Cousin," *Fulton Democrat,* Nov. 25, 1965.

25. Tri-State Natural Weather Association, *Cloud Seeding: The Science of Fraud and Deceit (A Criminally Conspired Complex)* (n.p., n.d.), 5.

26. National Research Council, *Weather and Climate Modification,* 107. See also R.A. Howard, J.E. Matheson, and D.W. North, "The Decision to Seed Hurricanes," *Science,* June 16, 1972, 1191–1202.

27. Budget constraints and a lack of hurricanes appropriate for seeding purposes led to the demise of Project Stormfury in 1972. See U.S. Department of Defense and Department of Commerce, NOAA, *Project Stormfury: Annual Report 1972* (Miami, Fla., 1973), 1–4.

28. Quoted in "Putting in the Weather Fix," *Nation,* Nov. 20, 1972. See also "'Crazy' Weather and the Food Crisis," a pamphlet available with other materials and fugitive papers of the Tri-State Natural Weather Association, many without date or place of publication, in the Records of the Pennsylvania Department of Agriculture.

29. Tri-State Natural Weather Association, "This Is the Internal Capture of the U.S," a broadside, Records of the Pennsylvania Department of Agriculture.

30. Tri-State Natural Weather Association, "Cloud Seeding: The Crime of the Century," advertisement, *Public Opinion* (Chambersburg, Pa.), Sept. 29, 1976.

31. Tri-State Natural Weather Association, *Cloud Seeding,* 1. The last point had support from ecologists who pointed out that hurricanes, for example, do serve a positive environmental function by moving water from one region to another. See, e.g., Frederick Sargent, "A Dangerous Game: Taming the Weather," *BAMS* 48 (1967): 452–458.

32. Quoted in "In the Clouds," *Philadelphia Inquirer,* Apr. 18, 1978.

33. Quotes are from ibid.

34. Richard Hofstadter, *The Paranoid Style in American Politics and Other Essays* (New York: Knopf, 1965), 39.

35. Quoted in "Charge Cloud Seeding Drove Tri-Stater's President out of Farming," *Gettysburg (Pa.) Times,* Mar. 31, 1977. See also Paul G. Hoke to Gov. Milton J. Shapp, July 29, 1974, Records of the Pennsylvania Department of Agriculture.

36. For more on the dominant "natural" view of calamity and the tendency to denounce alternative views implicating government, science, or business as conspiracy theories, see Kenneth Hewitt, "The Idea of Calamity in a Technocratic Age," in *Interpretations of Calamity: From the Viewpoint of Human Ecology,* ed. Kenneth Hewitt (Boston: Allen and Unwin, 1983), 17. Noam Chomsky once pointed out that "conspiracy theory has the approximate content of 'fuck you.'" "Noam Chomsky," Interview, *Cleveland Free Times,* Mar. 10–16, 1999, 10.

37. For the response of the natural-weather people to the 1972 flood, see, e.g., Tri-State Natural Weather Association, *Cloud Seeding: The Technology of Fraud and*

Deceit (The Anatomy of Murder) (n.p., n.d.), 17–18, Records of the Pennsylvania Department of Agriculture.

38. Perry H. Rahn, "Lessons Learned from the June 9, 1972, Flood in Rapid City, South Dakota," *Bulletin of the Association of Engineering Geologists* 12 (1975): 84.

39. *Black Hills Flood of June 9, 1972*, Natural Disaster Survey Report no. 72-1, (Rockville, Md.: Department of Commerce, NOAA, 1972), 1.

40. Figures are from "Flood Facts at a Glance," *Rapid City (S.Dak.) Journal*, May 17, 1992.

41. The following details of the seeding flights are from Pierre St.-Amand, Ray J. Davis, and Robert D. Elliott, "Report on Rapid City Flood of June 9, 1973," *Journal of Weather Modification* 5 (1973): 332–333. The article title gives the wrong year for the Rapid City flood.

42. Max Horkheimer, *Eclipse of Reason* (New York: Oxford Univ. Press, 1947), 97.

43. News release from Governor Kneip's Emergency Office, June 13, 1972, in Richard Schleusener Papers, Devereaux Library, South Dakota School of Mines and Technology, Rapid City, S.Dak.

44. Quotations are from, Memorandum, R.A. Schleusener to Governor Kneip and Mayor Barnett, June 13, 1972, Schleusener Papers.

45. Quoted in "Seeding Said Not Contributor to Flooding," *Rapid City Journal*, June 13, 1972.

46. Fred W. Decker to Sen. George McGovern, June 14, 1972, Schleusener Papers.

47. R.A. Schleusener, Memorandum for the Record on Meeting, June 19, 1972, Schleusener Papers.

48. Barbara C. Farhar and Julia Mewes, *Social Acceptance of Weather Modification: The Emergent South Dakota Controversy*, Institute of Behavioral Science, Program on Technology, Environment, and Man, monograph no. 23, (Boulder: Univ. of Colorado, 1976), 5.

49. Quotation is from St.-Amand's prepared statement in Senate Committee on Foreign Relations, *Weather Modification*, 93d Cong., 2d sess., 1974, Committee Print, 43. For details of the U.S. government's weather modification activities in Vietnam, see ibid., 87–123.

50. On the Yuba City flood, see Dean E. Mann, "The Yuba City Episode in Weather Modification," in *Legal and Scientific Uncertainties of Weather Modification*, ed. William A. Thomas (Durham, N.C.: Duke Univ. Press, 1977), 100–113.

51. See David Howell, "It Was an 'Act of God'—With a Few Grains of Salt," *Environmental Action*, May 12, 1973, 4–5.

52. In the 1960s, Schleusener headed up a committee that, in his words, sought "to counteract adverse publicity and to take positive steps to present favorable publicity to the public concerning weather modification." See Memorandum, Richard A. Schleusener to members of Weather Control Research Association, Mar. 15, 1966, Schleusener Papers.

53. See Merlin C. William's letter accompanying "Report on the Rapid City Flood of 9 June 1972," reprinted in House Committee on Public Works, *South Dakota Flood Disaster*, 92d Cong., 2d sess., 1972, Committee Print, 80.

54. Merlin C. Williams to Pierre St.-Amand, July 5, 1972, Administrative Files, box 1, file 5.3.1, Department of Natural Resources Development, Division of Weather Modification, South Dakota State Archives, Pierre, S.Dak.

55. H. Peter Metzger, "Did Rainmakers Swell Rapid City Deluge?" *Denver Post*, Sept. 24, 1972. The sentence in question reads, "While it cannot be shown that the seeding activities of June 9 augmented or diminished the storm and our reasoned conclusion is that the storm was not affected in a way that could have resulted in greater severity, we will comment on the practices employed." St.-Amand, Davis, and Elliott, "Report on Rapid City Flood," 336.

56. "South Dakota Flood—Was It Man-Made?" *Los Angeles Times*, Nov. 26, 1972.

57. "Gov't. Weather Tampering Is Causing World Floods," *National Tattler*, Dec. 24, 1972.

58. "Hysteria Shouldn't Curtail Scientific Weather Modification," *Rapid City Journal*, Dec. 28, 1972.

59. "Cloud Seeding at Rapid City: A Dissenting View," Letter to the Editor, *BAMS* 54 (1973): 676.

60. Richard A. Schleusener to Pierre St.-Amand, Jan. 3, 1973, Schleusener Papers.

61. Merlin C. Williams to director, Atomic Energy Commission, Nov. 6, 1972, ibid.

62. R.D. Elliott, "Review of 'Cloud Seeding at Rapid City' by Jack W. Reed," Nov. 1972, ibid. "The suggestion of holding off on cloud seeding until more is known about it," Elliott concluded, "is no more plausible than the suggestion of eliminating the automobile as a means of transportation until its effects on the environment are fully understood."

63. As St.-Amand saw it, the people of South Dakota had decided to support cloud seeding, which was true at least in regard to the state-sponsored seeding done under the auspices of the Weather Control Commission. But the experiment in question, Project Cloud Catcher, was not voted on. That did not stop St.-Amand from pointing out that "the public did however approve indirectly in that the Congress, recognizing a possible universal benefit from increased precipitation, allotted money for the work. Moreover, the people of South Dakota approved in that they, through legislative process, decided to let the South Dakota School of Mines and Technology engage in weather modification research activities." Pierre St.-Amand to Kenneth C. Spengler (editor, *BAMS*), Dec. 27, 1972, ibid.

64. My description of the following episode is from "Should We Try to Change Our Weather," *Denver Post*, May 6, 1973; and Luther J. Carter, "Weather Modification: Colorado Heeds Voters in Valley Dispute," *Science*, June 29, 1973, 1347–1350.

65. Farhar and Mewes, *Social Acceptance of Weather Modification*, 50.

66. "Haze Hangs over Smog Plan," *Denver Post*, Mar. 24, 1968.

67. Francis W. Bosco to E. Glasgo, June 1, 1973, Schleusener Papers.

68. Mrs. Adolph Hermann to Marion Bruce, Sept. 1, 1968, ibid.

69. Gertrude Milton Walcher, "How To Kill A Beautiful World" (Colorado Springs: Beautiful World Association, 1970), 1, 6, a copy of which can be found in the Schleusener Papers.

70. Ibid., 1.

SEVEN · FORECASTING AT THE FAIR WEATHER SERVICE

1. Quoted in "Ala. Town Still Feels Tornado's Force a Year Ago," *Philadelphia Inquirer*, Apr. 7, 1995.

2. "A Farrakhan Supporter Defends the Group America Loves to Hate," *Washington Post*, May 8, 1994.

3. Quoted in "Congregation Opens New Chapter with Church's Dedication," *Ledger-Enquirer* (Columbus, Ga.), July 15, 1996.

4. Two years before the tornado calamity, the agency tried to raise enough funds for new sirens in Piedmont but failed. See "Community Hit by Tornado to Get Money for Warning Sirens," *Ledger-Enquirer*, July 4, 1995. It is also interesting to make note of a 1987 calamity that bore an uncanny resemblance to the Piedmont tragedy. In the spring of that year, a tornado leveled the Hispanic community of Saragosa in Reeves County, Texas. Thirty people died and 121 were injured (out of a total population of 428). Twenty-two of the deaths occurred when the town's community center was struck during a graduation ceremony for Head Start. For the people in the hall there was no warning, save when a man, who knew of the tornado, rushed in to pick up his child. No sirens or public warning system existed in Saragosa. According to one respondent to a postdisaster survey, "Saragosa had been systematically ignored by public agencies." The county was also beyond the reach of NOAA Weather Radio at the time, though indirect warning was provided on the major television stations. See Benigno E. Aguirre, "The Lack of Warnings Before the Saragosa Tornado," *International Journal of Mass Emergencies and Disasters* 6 (1988): 66, 67, 70; and Benigno E. Aguirre et al., *Saragosa, Texas, Tornado, May 22, 1987: An Evaluation of the Warning System* (Washington, D.C.: National Academy Press, 1991), 30.

5. Quoted in "Ala. Town Still Feels Tornado's Force."

6. See "Community Hit by Tornado" and "Alabama Briefs," *Ledger-Enquirer*, Dec. 7, 1995.

7. Quotations are from "Weather Service Launches Bid to Arm Americans with Radios," *Gannett News Service*, Sept. 27, 1994.

8. Joseph G. Galway, "Early Severe Thunderstorm Forecasting and Research by the United States Weather Bureau," *Weather and Forecasting* 7 (1992): 565. Finley was the nineteenth century's most dedicated student of these phenomena, which in his lifetime caused increasing death and destruction as Americans forged their way across the continent. "Formerly, these violent meteors left no mark upon the treeless and uninhabited prairie," he wrote in 1887, ". . . but now the farm-house and the village dot the plain, and the hardy laborer has forced his way with his family into the depths of the forest." John P. Finley, *Tornadoes: What They Are and How to Observe Them; With Practical Suggestions for the Protection of Life and Property* (New York: Insurance Monitor, 1887), 9.

9. Galway, "Severe Thunderstorm Forecasting," 565, 568, 569.

10. Donald R. Whitnah, *A History of the United States Weather Bureau* (Urbana: Univ. of Illinois Press, 1961), 216.

11. Galway, "Severe Thunderstorm Forecasting," 571–572, 578.

12. See statement of Richard Hallgren in House Committee on Interstate and Foreign Commerce, *National Weather Warning System*, 94th Cong., 2d sess., 1976, Committee Print, 46.

13. Quoted in "Disasters," *Time*, Apr. 23, 1965, 29.

14. Quoted in "Man vs. Nature," *U.S. News & World Report*, Apr. 26, 1965, 52. On the role the tornado played in helping fuel early weather-radio efforts, see House Committee on Interstate and Foreign Commerce, *Weather Modification and Early-Warning Systems*, 93d Cong., 2d sess., 1974, Committee Print, 50.

15. House Committee on Interstate and Foreign Commerce, *National Weather Warning System*, 46. The figure on the United States' share of the world's tornadoes is from Mark Monmonier, *Cartographies of Danger: Mapping Hazards in America* (Chicago: Univ. of Chicago Press, 1997), 88.

16. See tables 4.3 and 4.4 in Thomas P. Grazulis, *Significant Tornadoes, 1680–1991* (St. Johnsbury, Vt.: Environmental Films, 1993), 37; and *The Widespread Tornado Outbreak of April 3–4, 1974*, Natural Disaster Survey Report no. 74-1 (Rockville, Md.: Department of Commerce, NOAA, 1974), v.

17. *Tornado Outbreak of April 3–4, 1974*, 2.

18. Quoted in "Tornado Alert System—Slow and Outdated," *Los Angeles Times*, June 13, 1974, reprinted in *Congressional Record*, 93d Cong., 2d sess., 1974, 120, pt. 21: 28522.

19. "A Disaster Warning System," *Congressional Record*, 94th Cong., 1st sess., 1975, 121, pt. 15: 19474. More generally, communications themselves at the NWS were a shambles, as Richard Hallgren, deputy director of the NWS, pointed out immediately after the 1974 calamity. "Our communications system isn't the best," he remarked. "It's really a hodgepodge…a guy in a weather station spends 25% to 30% of his time tearing off teletype copy." Hallgren said that he had long wondered "why the weather service was so cheap, why it had to nickel-and-dime it across the country." Quoted in "Tornado Alert System."

20. Quoted in "Tornadoes Striking More Often, with the South a Major Target," *Los Angeles Times*, June 10, 1974, reprinted in *Congressional Record*, 93d Cong., 2d sess., 1974, 120, pt. 21: 28522.

21. "This is a big workload. Clearly, we do not have sufficient manpower in most of our offices to meet this need," Hallgren told Congress in 1976. See House Committee on Interstate and Foreign Commerce, *National Weather Warning System*, 47.

22. Senate Committee on Appropriations, *Departments of State, Justice, and Commerce, the Judiciary, and Related Agencies Appropriations for Fiscal Year 1975*, 93d Cong., 2d sess., 1974, 850.

23. These numbers are from Richard L. Worsnop, "Progress in Weather Forecasting," *Congressional Quarterly's Editorial Research Reports*, June 15, 1990, 335.

24. *Big Thompson Canyon Flash Flood of July 31–August 1, 1976*, Natural Disaster Survey Report no. 76-1 (Rockville, Md.: Department of Commerce, NOAA, 1976), 1, 38; "Now, There's Nothing There," *Time*, Aug. 16, 1976, 23. See also the critique in "A Report on the Big Thompson Flood," *Four Winds*, June 1977, 5–7. *Four Winds* is the newsletter of the NWSEO.

25. *Johnstown, Pennsylvania, Flash Flood of July 19–20, 1977*, Natural Disaster Survey Report no. 77-1 (Rockville, Md.: Department of Commerce, NOAA, 1977), 1, 12. Quotation is from the undated report by Dave Leherr and Stuart Brown on the flood in the *Pittsburgh Post-Gazette* reprinted in the *Congressional Record*, 95th Cong., 1st sess., 1977, 123, pt. 24: 30483.

26. *Black Hills Flood of June 9, 1972*, Natural Disaster Survey Report no. 72-1, (Rockville, Md.: Department of Commerce, NOAA, 1972), 17–19. Apart from staffing problems, the floods also exposed the inadequacies of the weather service's information network. Critical data on rainfall and river levels simply were not available. Of course, accurate forecasts and warnings depend on good, reliable information, something denied the weather office in Rapid City during the 1972 flood because of too few surface observation stations. The spacing between stations in the states near South Dakota averaged between 100 and 200 miles. But 30 to 60 miles of distance, requiring the installation of nearly nine times as many weather observation stations, was needed to adequately forecast the kind of severe weather that Rapid City experienced during the flash flood. See ibid., 5.

    Better up-to-the-minute reports on rainfall in the Johnstown area also might have helped to reduce the death toll in the 1977 flash flood. Edward Epstein, who chaired NOAA's investigation into the disaster, explained that the weather service simply did not get a warning out in time to save lives. "Could lives have been saved?" Epstein asked rhetorically. "The answer has to be yes, if—and this is a big if—there were flood warning programs in effect and if people had responded to them." Epstein pointed out that some people in mobile homes had died after 3 A.M., that is, after it had been raining for hours. "With a total system of warning, communications and preparedness, lives there would have been saved." Quotations are from the report by Leherr and Brown, *Pittsburgh Post-Gazette*, reprinted in the *Congressional Record*, 30483.

27. *Big Thompson Canyon Flash Flood*, 38–39, 40, quotation from p. 38.

28. *Johnstown, Pennsylvania, Flash Flood*, 56, 58.

29. Figures are from table 3 in National Research Council, *Technological and Scientific Opportunities for Improved Weather and Hydrological Services in the Coming Decade* (Washington, D.C.: National Academy of Sciences, 1980), 72.

30. The study, done by J. Philip Bohart, David M. Harrington, and Norman L. Peterson and titled "Manpower Requirements and the Utilization of Manpower Resources Within the Weather Bureau, ESSA, DOC," is reprinted in House Committee on Science and Technology, *H.R. 13715—National Weather Service Act of 1978*, 95th Cong., 2d sess., 1978, Committee Print, 80–138. Quotation is from p. 80. In fact, between 1967 and 1977, Congress actually did authorize sufficient funds to hire 879 new weather service personnel. But what Congress gave, the Office of Management and Budget, the Department of Commerce, and NOAA conspired to take away. In order to meet employment ceilings set for the Department of Commerce, the NWS had to reduce the number of positions authorized by Congress by a figure of 839, explaining why the total personnel in the weather service remained unchanged during the decade.

Just how much staffing shortages affected the ability to provide adequate weather service was made amply clear in the 1970s at the Huron, South Dakota, station. The office was officially charged with issuing severe weather warnings in 20 counties. But because the stations at Aberdeen and Sioux Falls either had no radar or were unable to continuously staff the technology, the Huron office was left looking after some 53 counties in the state. The staffing shortage at the Huron office forced workers to rotate through shifts on a weekly basis, an onerous duty. Added to this was the burden of 15-hour shifts, required during severe weather in order to ensure public safety. To make matters worse, the lack of enough employees with the proper radar training meant that for 40 percent of the time between 10 P.M. and 4 A.M., only one employee staffed the entire weather operation—an employee who may well have not been fully capable of interpreting the radar. Here was a prescription for disaster that seemed to mock the fact that the worst flash floods of the 1970s occurred toward night. For more on the problems in South Dakota, see Senate Committee on Appropriations, *Departments of State, Justice, and Commerce, the Judiciary, and Related Agencies Appropriations for Fiscal Year 1978*, 95th Cong., 1st sess., 1977, Committee Print, 235–237, 244–246.

31. See Pearson's testimony in *H.R. 13715—National Weather Service Act of 1978*, 683–684.

32. See White's testimony in ibid., 196, 199.

33. The facts in the following discussion are from *Brown v. United States*, 599 F. Supp. 877 (D. Mass. 1984).

34. Ibid., 883–890.

35. *Brown v. United States*, 790 F.2d 199 (1st Cir. 1986).

36. The Reagan administration's philosophy of privatization in the name of reducing the size of government brought the NWS to the brink of extinction, only to see it rescued by Congress. The most publicized privatization scheme was the Reagan administration's attempt in 1983 to sell off parts of the NWS to private companies, starting with the government's system of weather satellites. Congress, however, led by Rep. James Scheuer of New York, who argued that the sale would result in second-rate weather services, not to mention a government-sponsored monopoly, was able to put a stop to this plan rather easily. Can you imagine what would happen if a city were to fail to pay its satellite bill prior to a major storm? the *Washington Post* asked. "Selling the Weather," *Washington Post*, Mar. 10, 1983, reprinted in *Congressional Record*, 98th Cong., 1st sess., 1983, 129, pt. 4: 5062.

Unsurprisingly, it took the Reagan administration just a month and a half after getting into office to propose cutting 38 weather offices. However, Congress managed to whittle this number down to 18. The following year the administration proposed shutting down 45 weather offices, but this move was later rescinded by the administration after a huge public outcry. See National Advisory Committee on Oceans and Atmosphere, *The Future of the Nation's Weather Services*, Special Report to the President and the Congress (Washington, D.C., 1982), 21. There was speculation that Richard Hallgren, the director of the NWS at the time, deliberately targeted those stations that would provoke the most public outcry.

37. The report suggests perhaps making FEMA responsible for warning preparedness, but no clear picture is offered of exactly how to do this. "National Weather Service: A Strategy and Organization Concept for the Future," (Report prepared by Booz-Allen and Hamilton for NOAA, June 24, 1983), p. 3; pt. 3, pp. 14–17.

38. National Advisory Committee on Oceans and Atmosphere, *Nation's Weather Services*, viii.

39. Yet, as the report noted, it was precisely "these interface links that we 'chip away at' when we reduce staff at weather stations and fail to provide them with the technological means to do the kind of dissemination and 'community advisory' job that is required." Ibid., 36.

40. Quoted in "Lives Lost or Transformed in Devastating Seconds," *Washington Post*, Mar. 30, 1984.

41. Quoted in "'Like the Hand of God,'" *Time*, Apr. 9, 1984, 18.

42. At the Raleigh office, there was at least one vacant position during the entire span between 1980 and 1984, and 40 percent of that time there were two or more vacancies. This information and other details on the 1984 disaster is from an unpublished Natural Disaster Survey Report titled "Carolina Tornadoes and Northeast Coastal Storm of March 28, 1984," 2, 17, available from the NWS, Office of Meteorology, Silver Spring, Md. The report notes that there were only 14 dedicated warning preparedness meteorologists nationwide—not exactly a field bursting with employment prospects. See ibid., 14.

43. See Park's testimony in House Committee on Science and Technology, *Field Briefing on Tornado Prediction and Preparedness in the Carolinas During the Severe Weather of March 28, 1984*, 98th Cong., 2d sess., 1984, Committee Print, 83, 84.

44. National Research Council, *Severe Storms: Prediction, Detection, and Warning* (Washington, D.C.: National Academy of Sciences, 1977), 56.

45. "Twisters Rip N. Carolina," *Washington Post*, Nov. 29, 1988.

46. The Natural Disaster Survey Report on the calamity titled "The North Carolina Tornadoes of November 28, 1988," is available in House Committee on Science, Space, and Technology, *Tornado Forecasting and Severe Storm Warning*, 100th Cong., 1st sess., 1989, Committee Print, 125, 138, 150, quotations from pp. 125, 138.

47. Quoted in "State Scheduled to Get Better Weather Radar," *MH*, June 6, 1990.

48. Apart from NEXRAD, there were other components to the multi-billion-dollar effort to modernize the weather service. These included a new satellite system, advanced data processing and communications, and automated surface observation that would allow machines to collect information on weather conditions, obviating the need to do this manually. For details on the modernization program, see U.S. General Accounting Office, *Weather Forecasting: Cost Growth and Delays in Billion-Dollar Weather Service Modernization*, GAO/IMTEC-92-12FS, Dec. 1991.

49. Dateline Transcript, Apr. 7, 1992, Records of the NWSEO, Washington, D.C. Strangely, Scheuer also seems not to have known about contributions made to his 1986 campaign by three people whom Sperry/Unisys had been using to make illegal congressional donations in order to woo business its way.

In 1987, production of NEXRAD began, but delivery of the units was delayed when Unisys got tangled up in a Justice Department inquiry into defense procurements. NOAA became more impatient with the delay and less certain that Unisys would actually be able to complete the contract. In 1991, NEXRAD deliveries were suspended. The delay, it need hardly be said, did not help those who were killed or injured when the radar went down in the 1988 North Carolina disaster. Eventually, however, NOAA and Unisys patched things up and delivery of the units resumed. See General Accounting Office, *Weather Forecasting*, 19–20.

50. Richard Hirn, attorney, NWSEO, interview with author, Cleveland, Apr. 11, 1997.

51. The number of offices changed over the years. These figures are for the office structure as it existed in 1975.

52. Quoted in "Modern Weather Warning System Criticized for Protection Gaps," *Durham (N.C.) Morning Herald*, June 18, 1989.

53. The plan advised equipping 113 WFOs with NEXRAD and turning the old WSOs—some 114 offices—into Local Weather Offices, which would be responsible for assisting communities in disaster preparedness and in developing spotter networks. See "Modernization and Restructuring of the National Weather Service," report prepared by Department of Commerce, NOAA Management Team, July 1984, pp. ES-5-6, available through a Freedom of Information Act Request, NWS, Silver Spring, Md.

54. See National Advisory Committee on Oceans and Atmosphere, *Nation's Weather Services*, 22. The authors of this report wrote: "Equivalent quality local weather services cannot, in most cases, be provided by consolidating weather forecasting functions at fewer weather stations, each of which would have a broader geographical responsibility. There is sufficient evidence about the relationship between accuracy of weather forecasts and service to community, and remoteness from the geographical location for which forecasts and the service are being provided to warrant concern."

55. Interview with Richard Hirn.

56. Paul Bieringer and Peter S. Ray, "A Comparison of Tornado Warning Lead Times with and without NEXRAD Doppler Radar," *Weather and Forecasting* 11 (1996): 47.

57. See Peter S. Ray, "Assessment of the National Radar Network," in House Committee on Science, Space, and Technology, *NEXRAD, Tornado Warnings, and National Weather Service Modernization*, 103d Cong., 2d sess., 1994, Committee Print, 281–282.

58. In *conventional* radar's reflectivity mode, however, coverage out to 125 miles was possible.

59. Quotations are from "Radar System Gets Mixed Forecast," *Huron (S.Dak.) Plainsman*, Apr. 20, 1993.

60. Quoted in "New Radar Could Leave Gaps in U.S. Weather Coverage," *Virginian-Pilot and the Ledger-Star* (Norfolk, Va.), June 11, 1989.

61. In fact, there were some holes in the umbrella. Parts of Alabama, Indiana, North Dakota, Pennsylvania, and Tennessee, it came out, were too far from planned

NEXRAD sites to benefit from the new technology. And because of scheduled office closings, the result would be degraded service in these areas. Under pressure from Congress, the National Research Council was hired to take a look at the purported gaps in the coverage. It confirmed the decline in service and advised the NWS to fix the problem, which it agreed to do by building more NEXRAD-equipped offices. Criticism of the NWS's modernization program was especially strong in Evansville, Indiana, an area well-known for its tornadoes. As initially outlined, the modernization plan called for closing the Evansville WSO and installing NEXRAD units in Paducah and Louisville, Kentucky, Indianapolis, St. Louis, and Springfield, Illinois, all of which were at least 90 nautical miles away from Evansville. As early as 1987 a study, commissioned by the Evansville Chamber of Commerce and conducted by Peter Ray and other qualified meteorological experts specializing in radar, showed that the area would not be adequately served by the distant NEXRAD installations. The NWS, however, was unmoved. In 1994, another study, again commissioned by the Evansville Chamber of Commerce and conducted by Ray, showed that coverage in northwestern Indiana as well as in northern Alabama and northwestern Ohio would not be adequate under the modernization plan. Again the NWS refused to budge. See House Committee on Science, Space, and Technology, *NEXRAD*, 64–66. See also National Research Council, *Toward a New National Weather Service—Assessment of NEXRAD Coverage and Associated Weather Services* (Washington, D.C.: National Academy Press, 1995). Eventually, however, the NWS, convinced of the problem, finally agreed to increase the number of NEXRAD-equipped offices to eliminate the gaps in coverage.

62. According to Richard Hirn, some people suspect that Leon Panetta, at the time the director of the Office of Management and Budget, may have played an important role in the planned relocation to Monterey. Panetta was born there and his family remained a major landowner in the area.

63. This quotation and the ones in the preceding paragraph are from "Critics Pelt Relocation of Weather Service," *SFE*, Mar. 7, 1993.

64. "A relocation to the distant site of Monterey will make these face-to-face meetings much more difficult," Medigovich explained. See William M. Medigovich to Nancy Pelosi, Mar. 26, 1993, Records of the NWSEO.

65. Quoted in "Critics Pelt Relocation."

66. Local Branch 4-30, Redwood City, Calif., "Why We Shouldn't Move to Monterey," p. 5, Records of the NWSEO.

67. The NWSEO brought suit in conjunction with the city of Harrisburg against the Department of Commerce in 1992. See Brief in Support of Plaintiffs' Motion for Partial Summary Judgment, or, in the Alternative, for a Preliminary Injunction, *City of Harrisburg* v. *Franklin*, Records of the NWSEO.

68. Quotations are from "Union Opposes Weather Service Relocation Plan," *Centre Daily Times* (State College, Pa.), Mar. 16, 1992.

69. One disturbing indication of this overreliance on technology to the detriment of human observation is apparent in the NWS's tornado-spotter program. Prior to

254 Notes to pages 170–71

the deployment of Doppler radar, meteorologists would often wait until a spotter reported the sighting of a tornado before issuing a warning. With the development of Doppler, however, the warnings are based on what the radar shows, information that the meteorologist then uses to decide whether to issue a warning. The problem is that the desire to not let the public down by failing to issue a warning for dangerous weather has led to a proliferation of warnings that have turned out wrong. In 1997, the NWS issued 2,592 tornado warnings and 2,022 were false alarms. The false alarms, in turn, have caused people to ignore tornado warnings. Meanwhile, there is some evidence that the NWS is allowing its spotter program to languish. Some meteorologists have complained that there are not enough spotter training booklets available and, indeed, in 1998 a spokesperson for the NWS admitted that there had been budget cuts in the spotter program. See "False Weather Warnings Blowin' in the Wind," *Times Union* (Albany, N.Y.), May 10, 1998.

70. Quoted in "Hurricane Planners Cry Foul over Cuts," *Times-Picayune* (New Orleans), Apr. 14, 1997.

71. Quoted in "National Hurricane Center Will Make Cuts," *Post and Courier* (Charleston, S.C.), Mar. 18, 1997.

72. Quoted in "Hurricane Center Cuts Too Costly, Experts Warn," *Palm Beach (Fla.) Post*, Mar. 28, 1997.

73. Figure on the death toll is from "Penny-Wise Cutbacks," *Sarasota (Fla.) Herald-Tribune*, Mar. 22, 1997.

74. "Forecasters Link Budget Cuts to Coast Guard Crew's Deaths," *Seattle Times*, Mar. 25, 1997.

75. Maggie Gross to Ted Bushner, n.d., electronic mail, Records of the NWSEO.

76. "National Weather Service Officials Blast Cuts," *Ledger-Enquirer*, Apr. 27, 1997.

77. Quoted in "Coast Guard Crew's Deaths."

78. Quoted in "Shaw Vows to Fight Hurricane Center Cuts," *MH*, Mar. 29, 1997.

79. Quoted in "4 Officials Protest Weather Service Cuts," *Los Angeles Times*, Mar. 25, 1997.

80. Quoted in "How Will We Weather This Unkind Cut?" *MH*, Mar. 23, 1997.

81. William M. Daley to Harold Rogers, Apr. 17, 1997, Records of the NWSEO.

82. Quoted in "Forecasters Brace for Budget Ax" *MH*, Mar. 25, 1997. In June 1997, Joe Friday, director of the NWS, was removed from his position and reassigned within NOAA. A study, headed by John Kelly, a retired brigadier general in the air force, subsequently found that the NWS proved inconsistent and unreliable in regard to its budget projections. See "Forecasters Weather Needed Change, Report Says," *Washington Post*, Oct. 24, 1997. Then, in February 1998, Kelly, who had a reputation for budget cutting forged during his years in the military, was made director of the NWS. McPherson said of Kelly, "I know in his heart he would like the Weather Service to be simpler, but the fact is the Weather Service is complicated." Quoted in "His Forecast Is for Change," *Washington Post*, May 5, 1998.

EIGHT · WHO PAYS?

1. "'It Felt Like 15 or 20 Tornadoes Had Caught Us,'" *Newsweek*, Sept. 1, 1969, 18.

2. For more on equity issues and natural disasters, see Raymond J. Burby, *Sharing Environmental Risks: How to Control Governments' Losses in Natural Disasters* (Boulder, Colo.: Westview Press, 1991), 41–45. This study examines data from 1980 to 1987 and shows that small jurisdictions were shortchanged on relief aid, though there is no evidence, in this study at least, that poorer areas were discriminated against.

3. James M. Gere and Haresh C. Shah, *Terra Non Firma: Understanding and Preparing for Earthquakes* (San Francisco: W.H. Freeman, 1984), 35–36, 46–47, 50.

4. This figure on congressional districts is derived from the table in House Committee on Public Works, *Disaster Relief Act of 1965*, 89th Cong., 2d sess., 1966, Committee Print, 90–93.

5. There were 128 enactments between 1800 and 1949. See Burby, *Sharing Environmental Risks*, 5.

6. Quoted in American Friends Service Committee, "In the Wake of Hurricane Camille: An Analysis of the Federal Response," Nov. 24, 1969, in Senate Committee on Public Works, *Federal Response to Hurricane Camille*, pt. 2, 91st Cong., 2d sess., 1970, Committee Print, 694.

7. For a good short history of federal disaster relief, see OEP, *Disaster Preparedness*, Report to Congress (Washington, D.C., 1972), 1: 167–173. See also Peter J. May, *Recovering from Catastrophes: Federal Disaster Relief Policy and Politics* (Westport, Conn.: Greenwood, 1985), 18–31.

   One unintended consequence of the increased generosity of the federal government's relief efforts, notes Rutherford Platt, is that states have been reluctant to pursue mitigation programs that would head off disaster before it strikes. In other words, state governments are likely to feel that the federal government will bail them out in the event of a natural disaster, and thus that mitigation is not money well spent. See Rutherford Platt, "Hazard Mitigation: Cornerstone or Grains of Sand?" *Natural Hazards Observer* 21 (1996): 11.

8. *Congressional Record*, 91st Cong., 1st sess., 1969, 115, pt. 6: 7643, 7645.

9. See May, *Recovering from Catastrophes*, 107.

10. This placed the storm among the top 10 most costly in U.S. history. See table 3a in Paul J. Hebert, Jerry D. Jarrell, and Max Mayfield, *The Deadliest, Costliest, and Most Intense United States Hurricanes of This Century (And Other Frequently Requested Hurricane Facts)*, NOAA Technical Memorandum NWS TPC-1 (Washington, D.C.: Department of Commerce, 1996), 8. Much of Camille's damage occurred in Mississippi, though the storm eventually headed north toward Virginia, where huge quantities of rain—some 25 inches in spots—caused massive flooding along the James River.

11. Quoted in "In the Wake of Camille," *New Republic*, Jan. 10, 1970, 9.

12. Senate Committee on Public Works, *Hurricane Camille*, pt. 2, 701. Nor did the federal government provide anything remotely resembling consumer or legal protection to victims of the disaster. The closest it came to providing legal-services funding was a $50,000 grant made by the Office of Economic Opportunity to the Mississippi

State Bar Association, presided over by Boyce Holleman, who rose to fame prose-
cuting civil rights workers in the mid-1960s as a district attorney, and who made no
secret of the fact that he would use the money not to help poor blacks recover from
the calamity but to aid middle-class whites as they dealt with insurance issues. In
truth, because both the federal and state government abdicated on consumer protection, there were plenty of such insurance problems to keep bar association attorneys busy. Some 600 shotgun adjusters flooded into Mississippi after the storm, no
doubt to the glee of Wall Street, which worried that the total losses from the storm
would soften the insurance industry's enthusiasm for municipal bonds unless claims
could be minimized. Preying on cash-starved victims, the adjusters were able to settle
close to 90 percent of the claims by early November, paying many of the insured
just 10 to 30 percent of the face value of their policies. As written, these policies did
not pay for water damage, only for destruction caused by wind, a legal fiction that
worked to the advantage of the insurance companies, which hired engineering firms
to document wind damage in order to further deprive policyholders of rightful settlements. See "In the Wake of Camille," *New Republic*, 9; and Senate Committee on
Public Works, *Hurricane Camille*, pt. 2, 696, 984, 1001–1002.

On the state level, Gov. John Bell Williams, a man not exactly known for his
racial tolerance, delivered an object lesson on how to go about restoring the normal
divisions of race and class. Williams appointed an emergency council made up of 10
men, all white, most of whom were bankers. The council set its sights very high, so
high, in fact, that it completely overlooked the need for any plan to deal with the
thousands of people lower down the social scale made homeless in the disaster, a
point openly admitted by Leo Seal, a Gulfport banker and council member. Instead,
the council busied itself with talk about transforming the Gulf Coast into a resort
replete with golf courses, marinas, and tony apartment buildings. None of this
seems to have bothered President Richard Nixon, who ordered all federal relief
efforts to be coordinated directly through the governor's newly formed council.
Only near the end of 1969, when Democrats in Congress led by senators Bayh and
Edmund Muskie planned hearings on the inadequacies of the federal response to
Camille, did the Nixon administration pressure Williams into opening up the coun-
cil to blacks. See "Gulf Coast 4 Months After Camille," *Delta Democrat-Times*
(Greenville, Miss.), Dec. 23, 1969, reprinted in Senate Committee on Public Works,
Hurricane Camille, pt. 2, 735; and Fred J. Russell, Memorandum for General
Lincoln, Jan. 12, 1970, Records of the OEP, Hurricane Camille Study Files, Record
Group 396, entry 1079, box 7, National Archives, College Park, Md.

13. Senate Committee on Public Works, *Hurricane Camille*, pt. 2, 842.
14. See Thompson's testimony in Senate Committee on Banking, Housing, and
    Urban Affairs, *SBA Administration of the Disaster Relief Program*, 92d Cong., 2d
    sess., 1972, Committee Print, 46, quotation from p. 54.
15. See the report by the American Friends Service Committee titled "The Agnes
    Disaster and the Federal Response," in Senate Committee on Public Works, *To
    Investigate the Adequacy and Effectiveness of Federal Disaster Relief Legislation*, pt. 3,
    93d Cong., 1st sess., 1973, Committee Print, 1679.

16. For more on the tax issue, see U.S. General Accounting Office, *Federal Disaster Assistance: What Should the Policy Be?* PAD-80-39, June 1980, 18. See also May, *Recovering from Catastrophes*, 142.

17. Quoted in "Relief Is the Real Disaster," *Nation*, July 2, 1977, 18. The discussion of the flood that follows is also based on House Committee on Small Business, *Federal Natural Disaster Assistance Programs*, 95th Cong., 1st sess., 1977, Committee Print, 322, 330, 332.

18. "With the damage present," concluded investigators sent by Congress to check up on how disaster relief was proceeding, "the application rate is much, much too low as is the disbursal rate." House Committee on Small Business, *Federal Natural Disaster Assistance Programs*, 326.

19. "Central Appalachia—The Forgotten Colony," in ibid., 110.

20. *Congressional Record*, 92d Cong., 2d sess., 1972, 118, pt. 19: 24617.

21. James N. Wood Jr., of the Southern Regional Council discusses the Florida disaster in Senate Committee on Public Works, *Hurricane Camille*, pt. 5, 2430. See also "Flood Compared to '47 Tragedy," *MH*, Mar. 28, 1970; and "Declare Disaster in Seven Counties, Kirk Asks U.S.," *MH*, Apr. 2, 1970.

22. Perhaps nowhere was this bias in favor of physical destruction at the expense of personal suffering more apparent than in the aftermath of the 1972 Rapid City flood. All kinds of federal aid flowed into the city to help with the reconstruction of roads, bridges, and other facilities destroyed during the disaster. But some months after the calamity, when it became clear that Rapid City was in the midst of a major mental health crisis brought on by the flood, federal support simply could not be mustered. Area mental health clinics reported an increase in the number of depressed patients seeking help. The state mental hospital noted a rise in admittances among people from the Rapid City area. But because the existing social and psychological service network did not have the staff or resources to cope with the marked rise in mental illness, some groups found it impossible to get adequate treatment for their emotional problems. Chief among those whose mental health needs went unaddressed were children, the elderly, and those lower down the socioeconomic scale, especially Native Americans, who suffered a disproportionate number of deaths in the flash flood tragedy (they made up just 5 percent of the population of Rapid City, but accounted for 14 percent of the deaths). And yet, over a year after the flood, well after the mental health crisis had become apparent, the federal government still had failed to provide money for outreach workers and clinical care. While it is true that the government was more successful in addressing the mental health needs of the victims of the San Fernando earthquake and Hurricane Agnes, it is still the case that the federal response to disaster continued to be biased in favor of countering the physical destruction of property at the expense of individual suffering, a response that tended to benefit the better off in society. On the mental health problems in Rapid City, see Senate Committee on Public Works, *To Investigate the Adequacy and Effectiveness of Federal Disaster Relief Legislation*, pt. 2, 291–301.

23. Instances in which human involvement played a major role typically resulted in the denial of aid. Some of the famous turndowns included: riot damage in Detroit (1967), a railroad car explosion in Illinois (1972), the collapse of a bridge in Tennessee (1974), a power outage in New York (1977), and oil spills in Massachusetts and Rhode Island (1977). Despite the attempt to limit relief to natural hazards, however, there were some instances in which acts of man managed to qualify. Of the ten such man-made events, eight occurred in the 1970s alone. Five of these events were dam breakings, including the breach of an industrial dam that gutted the community of Buffalo Creek in West Virginia (1972), and the Grand Teton Dam collapse (1976), which destroyed over 100,000 acres of farmland in the West. Three of the exceptions involved chemical contamination, two declarations alone for Love Canal. There was one chlorine barge sinking in 1962, and the Cuban refugee crisis in 1980. Still, when you consider that over 600 disaster declarations were made in three decades and that only a handful of cases involved events with an overwhelming human component, it seems clear that disaster relief was designed mainly to deal with natural calamity. The best historical treatment of this issue is contained in a study published by the Congressional Research Service and prepared by Clark F. Norton titled "'Other Catastrophe' Statutory Authority for Major Disaster Declarations," reprinted in *Congressional Record*, 96th Cong., 2d sess., 1980, 126, pt. 21: 27644–27666.

24. See, e.g., the remarks of Sen. Quentin Burdick of North Dakota in ibid., 27662. See also Senate Committee on Environment and Public Works, *Disaster Relief Act Amendment of 1980*, 96th Cong., 2d sess., 1980, S. Rept. no. 96-891.

25. See, e.g., Senate Committee on Environment and Public Works, *Disaster Relief Act Amendments of 1983*, 98th Cong., 1st sess., 1984, Committee Print.

26. 42 U.S.C. § 5122. On the debate over the amendments to the law that went into effect in 1988, see *Congressional Record*, 100th Cong., 2d sess., 1988, 134, no. 33: H938–965; and Senate Committee on Environment and Public Works, *Disaster Relief Act Amendments of 1988*, 100th Cong., 2d sess., 1988.

27. Peter Linebaugh, "Gruesome Gertie at the Buckle of the Bible Belt," *New Left Review*, no. 209 (1995): 16.

28. *Congressional Record*, 102d Cong., 2d sess., 1992, 138, no. 67: H3267.

29. Perhaps the only person more upset than Emerson about the Los Angeles riot was Sebastian Garafalo, who happened to be the mayor of Middletown, Connecticut, when spring flooding that very same year caused heavy destruction in the state, though not heavy enough to warrant a federal disaster declaration. "If a natural disaster can't get you funds, and a man-made one can get funds, it really tells a hell of a story," he remarked. Quoted in "Bush Denies Flood Disaster Aid for State," *Hartford Courant*, July 2, 1992.

30. *Congressional Record*, 102d Cong., 2d sess., 1992, 138, no. 67: H3261–3262.

31. Quoted in "Rain Thwarts Effort to Plug Chicago Tunnel Leak," *NYT*, Apr. 16, 1992.

32. See William Cronon's op-ed essay, "Mud, Memory and the Loop," *NYT*, May 2, 1992.

33. At the time of the 1993 midwestern floods, Sen. Dennis DeConcini of Arizona tried to get disaster relief for a Native American reservation in his state, arguing that economic deprivation there—caused in part by a freeze on new construction imposed in 1966 by Robert Bennett, the U.S. Indian Commissioner—qualified as a natural disaster. See *Congressional Record*, 103d Cong., 1st sess., 1993, 139, no. 112, pt. 2: S10329–10333.

34. For the conservative view, consider these thoughts by Marvin Kosters of the right-wing American Enterprise Institute: "I tend to have a soft spot for disaster," he remarked during the 1993 midwestern floods. "There's more of a legitimate federal role for these unexpected, unusual contingencies than there is for lots of programs we have going on a regular basis." Quoted in "Experts Urge More Disaster Preparedness," *SLPD*, July 17, 1993.

35. Quoted in the epigraph to Robert Scheer, *With Enough Shovels: Reagan, Bush, and Nuclear War* (New York: Random House, 1982).

36. House Committee on Government Operations, *Managing the Federal Government: A Decade of Decline*, 102d Cong., 2d sess., 1993, Committee Print, 125.

37. "It's Politics over Skill at Disaster Relief Agency," *Atlanta Constitution*, Sept. 12, 1992.

38. A copy of Giuffrida's thesis is reprinted in Senate Committee on Governmental Affairs, *Nomination of Louis O. Giuffrida*, 97th Cong., 1st sess., 1981, Committee Print, 34–83.

39. Daniel Franklin, "The FEMA Phoenix," *Washington Monthly*, July/Aug. 1995, 39.

40. These presidential appointees required, of course, Senate confirmation. See House Committee on Government Operations, *Managing the Federal Government*, 125.

41. "FEMA Focusing on Nuclear War, Not Natural Disasters," *Houston Chronicle*, Feb. 22, 1993.

42. Quoted in "After the Deluge," *National Journal*, Apr. 18, 1987, reprinted in House Committee on Public Works and Transportation, *Reauthorization of the Federal Disaster Relief Program*, 100th Cong., 1st sess., 1987, Committee Print, 172.

43. Quoted in "Lag in US Aid Angers Hugo Victims," *Washington Post*, Oct. 4, 1989. See also "Rural Areas Plead for Aid," *NC*, Sept. 27, 1989; House Committee on Government Operations, *Managing the Federal Government*, 126; "Effort Mounted to Aid Forgotten Hugo Victims," *Charlotte Observer*, July 4, 1990; and "FEMA Halts Funds for Hugo Outreach," *NC*, Feb. 8, 1991.

44. See Perata's testimony in House Committee on Banking, Finance, and Urban Affairs, *Housing Needs in Earthquake Disaster Areas*, 101st Cong., 2d sess., 1990, Committee Print, 72.

45. Quoted in "Many Still Homeless Long After Earthquakes," *USA Today*, May 21, 1990.

46. House Committee on Banking, Finance, and Urban Affairs, *Housing Needs*, 7, 17, 60, 169, 170–171. Nearly two-thirds of the housing units destroyed or severely compromised in the earthquake were multifamily rental units. See Mary C. Comerio, *Disaster Hits Home: New Policy for Urban Housing Recovery* (Berkeley: Univ. of California Press, 1998), 66.

47. House Committee on Banking, Finance, and Urban Affairs, *Housing Needs*, 35.

48. Ibid., 88.

49. See the testimony of Ilene Weinreb of the Oakland Housing Organization in ibid., 96.

50. Ibid., 90. However, Steve Ronfeldt of the Legal Aid Society of Alameda County said that of the roughly 2,500 people in SROs, some 2,100 failed to receive housing assistance.

51. Quoted in ibid., 90.

52. Quoted in "Quake Relief Ordered for Residential Hotels," *SFC,* Apr. 19, 1990. See also "US Agency Trying to Modify Quake-Relief Plan, Critics Say," *SFC,* Mar. 28, 1990.

53. Quoted in "Reform of Disaster Aid Agency Sought," *Los Angeles Times,* Oct. 28, 1992.

54. "FEMA Relief Efforts After 3 Crises Criticized," *MH,* Oct. 28, 1992.

55. Quoted in "The Hidden Toll," *Hawaii Business,* Sept. 1993.

56. Quoted in "Bay Area's Housing Situation Still Shaken One Year After Quake," *San Diego Union-Tribune,* Oct. 21, 1990.

57. Franklin, "The FEMA Phoenix," 41.

58. See table 5.1 in *Federal Disaster Assistance: Report of the Senate Task Force on Funding Disaster Relief,* 104th Cong., 1st sess., 1995, S. Doc. 104-4, 77. The table provides figures through 1994. In 1995, a supplemental appropriation for disaster in excess of $6 billion was passed.

59. "Durenberger, Penny Pose Question on Relief Aid," *Star Tribune* (Minneapolis, Minn.), Sept. 17, 1992.

60. *Congressional Record,* 102d Cong., 2d sess., 1992, 138, no. 122: H8226.

61. "Deficit Concerns Stall Flood Aid Bill," *Los Angeles Times,* July 23, 1993.

62. *Congressional Record,* 103d Cong., 1st sess., 1993, 139, no. 112: S10311.

63. Quoted in "The Real Disaster: How the L.A. Earthquake Added to the Federal Deficit," Editorial, *Pittsburgh Post-Gazette,* Feb. 23, 1994.

64. *Congressional Quarterly Almanac: 103rd Congress 2nd Session, 1994* (Washington, D.C.: Congressional Quarterly, 1995), 50: 548.

65. *Congressional Quarterly Almanac: 104th Congress 1st Session, 1995* (Washington, D.C.: Congressional Quarterly, 1996), 51: 11–105.

66. Deborah A. Frank et al., "Seasonal Variation in Weight-for-Age in a Pediatric Emergency Room," *Public Health Reports* 111 (1996): 366–371.

67. Steven Whitman et al., "Mortality in Chicago Attributed to the July 1995 Heat Wave," *American Journal of Public Health* 87 (1997): 1515–1518; "Tragedy in Chicago," *Weatherwise* 49 (1996): 18.

68. "Non-Hispanic African Americans were 1.9 times more likely to die from the heat as non-Hispanic whites, and this difference was evident in all ages over 55." See Stanley A. Changnon, Kenneth E. Kunkel, and Beth C. Reinke, "Impacts and Responses to the 1995 Heat Wave: A Call to Action," *BAMS* 77 (1996): 1501. Another study of the 1995 Chicago heat wave observed that "those at greatest risk of dying from the heat were people with medical illnesses who were socially iso-

lated and did not have access to air conditioning." See Jan C. Semenza et al., "Heat-Related Deaths During the July 1995 Heat Wave in Chicago," *New England Journal of Medicine* 335 (1996): 84. See also T. Stephen Jones et al., "Morbidity and Mortality Associated with the July 1980 Heat Wave in St. Louis and Kansas City, Mo.," *Journal of the American Medical Association* 247 (1982): 3327–3331, which talks about other data on the role of race and poverty in these kinds of events.

69. Chicago's medical examiner Edmund Donoghue remarked: "The cause of this is an act of God. This is a disaster unlike any we've ever seen in Chicago and I don't think any human being is responsible for this." Quoted in "Why Did So Many Die in Chicago?" *Cincinnati Post*, July 20, 1995. On Chicago's lack of preparation for extreme heat, see *July 1995 Heat Wave*, Natural Disaster Survey Report, no number, (Silver Spring, Md.: Department of Commerce, NOAA, 1995), viii–ix.

70. Changnon, Kunkel, and Reinke, "1995 Heat Wave," 1499. During the August 1988 Chicago heat wave, there were 232 so-called excess deaths, that is, the number of deaths above the average number of deaths for August in years past. Excess deaths during heat waves are considered by some experts to be the most accurate measure of heat-related deaths.

71. F.P. Ellis, "Mortality from Heat Illness and Heat-Aggravated Illness in the United States," *Environmental Research* 5 (1972): 51.

72. Quoted in "Heat Toll May Top 15,000 Killed," *Arizona Republic*, Aug. 17, 1988. In response to the monstrous death toll in Chicago, the National Association of Medical Examiners, led by Chicago's Edmund Donoghue, did issue a recommended definition for heat death to help standardize the figures. See Edmund R. Donoghue et al., "Criteria for the Diagnosis of Heat-Related Deaths: National Association of Medical Examiners," *American Journal of Forensic Medicine and Pathology* 18 (1997): 11–14.

73. "Taking the Heat," *American Medical News*, Aug. 21, 1995. In the wake of the calamity, Chicago's mayor convened a commission on extreme weather, and the city has taken steps to improve its watch and warning system, as well as its mitigation program. Still, Dr. John Wilhelm of the Chicago Department of Public Health worries that the mild summers since 1995 have helped to expand the pool of people susceptible to heat-related death. Meanwhile, Laurence S. Kalkstein, an extreme-weather expert at the University of Delaware, reports that it is New York City that is likely to eventually experience real trouble. Kalkstein reports that excess deaths in New York City during the summertime have long been a problem, yet city officials seem oblivious to the issue. Dr. John Wilhelm, deputy commissioner for medical affairs, Chicago Department of Public Health, telephone interview with author, Mar. 30, 1999; Dr. Laurence S. Kalkstein, associate director, Center for Climatic Research, University of Delaware, telephone interview with author, Mar. 26, 1999.

74. "Heat Wave's Final Chapter Is Cold, Lonely," *Los Angeles Times*, Aug. 26, 1995; "Emergency Funds Provided to Help Combat Heat Wave," *Charleston Gazette*, July 22, 1995.

75. Carol Moseley-Braun, Paul Wellstone, and Edward Kennedy were the only senators who paid the heat wave any attention at the time. See *Congressional Record*, 104th Cong., 1st sess., 1995, 141, no. 118: S10420–10422, S10457–10460.

76. The letter was read into the *Congressional Record*, 104th Cong., 1st sess., 1995, 141, no. 26: S2382.

77. *Congressional Quarterly Almanac, 1995*, 11–96.

78. In 1997, Congress passed a $8.9-billion emergency spending bill that provided money to victims of the floods in the Upper Midwest and for Pentagon missions in Bosnia and Iraq. The bill included recissions in unspent military funding and federal transportation funds as well as in the Section 8 housing program. Later in the year, Congress restored the housing money through a special exemption. In 1998, Congress passed a $5.6 billion supplemental spending bill that included $2.3 billion in cuts in Section 8 housing funds. The money for housing would have to be replaced with cuts to other domestic spending programs.

79. Mike Davis, *Ecology of Fear: Los Angeles and the Imagination of Disaster* (New York: Metropolitan Books, 1998), 51. In the wake of Hurricane Mitch in 1998, Republicans in both the House and Senate held up almost $1 billion in aid to Central America, demanding an equal amount be cut from domestic programs such as welfare assistance and food stamps. See "Budget Politics Is Delaying Relief for Central America," *NYT*, Mar. 5, 1999.

EPILOGUE · REMEMBERING McKINNEYSBURG

1. The following description of the flood is based on "Rivers Unleashed: Flood of '97," *Cincinnati Enquirer*, July 27, 1997; "Forecasters Missed Licking River's Crest at Falmouth," *Courier-Journal* (Louisville), Mar. 7, 1997; "Rain and Ruin," *Courier-Journal*, Mar. 8, 1997; "Rivers Unleashed: Flood of '97," *Cincinnati Enquirer*, Mar. 10, 1997; and "Flooding Victims Put Lives at Risk to Guard Homes," *Chattanooga Free Press*, Mar. 9, 1997.

2. The NWS's assessment of the disaster notes, "The lack of telemetered stream level information delayed critical information from being relayed to NWS forecasters and hydrologists. However, the impact on the timeliness and accuracy of the river stage forecast at Falmouth is not exactly known as the river forecast system could not be reproduced during the service assessment." See *Ohio River Valley Flood of March 1997*, Office of Hydrology Service Assessment (Silver Spring, Md.: Department of Commerce, NOAA, 1998), B-6.

3. See "Rivers Unleashed: Flood of '97," *Cincinnati Enquirer*, Apr. 7, 1997.

4. Quoted in "Hydrologist Warned of 'Serious Consequences' If Water-Depth Gauges Were Removed," *Gannett News Service*, Mar. 12, 1997.

5. In March 1998, after a long struggle to find the money for the project, the Kentucky Department of Disaster and Emergency Services agreed to pay for the installation of a new stream gauge in Pendleton County. See "Rain and Ruin: One Year Later," *Courier-Journal*, Mar. 4, 1998.

6. Ibid.

7. National Advisory Committee on Oceans and Atmosphere, *The Nation's River and Flood Forecasting and Warning Service,* Special Report to the President and the Congress, (Washington, D.C., 1983), 30.

8. Eaton's views were expressed in testimony before Congress. See House Committee on Appropriations, *Department of the Interior and Related Agencies Appropriations for 1998,* pt. 8, 105th Cong., 1st sess., 1997, Committee Print, 37.

9. According to the 1990 census, one in four people in the city has income below the poverty level. See table 9 in Department of Commerce, Bureau of the Census, *1990 Census of Population and Housing: Summary Social, Economic, and Housing Characteristics: Kentucky* (Washington, D.C., 1992), 109.

10. One study has noted that the number of gauge stations 50 years or older has not been growing at the rate one would expect, suggesting a falloff in the older stations since 1970. The study indicated that in 1960 there were roughly 3,500 gauges at least 20 years old. By 1990, one would have expected that roughly the same number of gauges would be celebrating their fiftieth anniversary. In fact, the actual number of gauges over 50 years old in operation was only slightly over 2,000. See fig. 9 in Kenneth L. Wahl, Wilbert O. Thomas Jr., and Robert M. Hirsch, "The Stream-Gaging Program of the U.S. Geological Survey," U.S. Geological Survey Circular no. 1123 (Reston, Va., 1995), 12.

11. Real-estate development in cities near Falmouth has increased the impervious ground cover, causing, in turn, an increase in runoff into the Licking River.

12. Quoted in "Awaiting Quake Aid, and Riot Aid Too," *NYT,* Jan. 27, 1994.

BIBLIOGRAPHY

....

MANUSCRIPT COLLECTIONS

Bacot, Julius Mott. Papers, 1886. South Carolina Historical Society, Charleston.

California Supreme Court. Legal papers in *California Wine Association* v. *Commercial Union Fire Insurance Company of New York* and *Pacific Heating and Ventilating Company* v. *Williamsburgh City Fire Insurance Company*, 1910. California State Archives, Sacramento.

Cantwell, L.E. Papers, 1886. South Carolina Historical Society.

Courtenay, William Ashmead. Correspondence, 1886. South Carolina Historical Society.

Floods in St. Louis and Vicinity. Scrapbooks. Missouri Historical Society, St. Louis.

National Weather Service Employees Organization. Papers. Washington, D.C.

San Francisco Chamber of Commerce. Papers, 1906–1907. California Historical Society, San Francisco.

San Francisco Earthquake and Fire. Papers, 1906–1908, 1919, 1956–1978. California Historical Society.

Schleusener, Richard. Papers. Devereaux Library. South Dakota School of Mines and Technology, Rapid City.

South Dakota Department of Natural Resources Development, Division of Weather Modification. Administrative Files. South Dakota State Archives, Pierre.

Tri-State Natural Weather Association. Papers. Pennsylvania Department of
   Agriculture, Harrisburg.
U.S. Office of Emergency Preparedness. Hurricane Camille Study Files, record group
   396. National Archives, College Park, Md.
Waterways Collection. St. Louis Mercantile Library, St. Louis, Mo.
Works Progress Administration. Papers. Federal Emergency Relief Administration.
   Central Files, record group 69.006.1. National Archives, Washington, D.C.

BOOKS AND ARTICLES

Aguirre, Benigno E. "The Lack of Warnings Before the Saragosa Tornado."
   International Journal of Mass Emergencies and Disasters 6 (1988): 65–74.
Aguirre, Benigno E., Walter A. Anderson, Sam Balandran, Brian E. Peters, and H. Max
   White. *Saragosa, Texas, Tornado, May 22, 1987: An Evaluation of the Warning System.*
   Washington, D.C.: National Academy Press, 1991.
Albala-Bertrand, J.M. *Political Economy of Large Natural Disasters: With Special
   Reference to Developing Countries.* Oxford: Clarendon Press, 1993.
American National Red Cross. *The West Indies Hurricane Disaster, September, 1928:
   Official Report of Relief Work in Porto Rico, the Virgin Islands and Florida.*
   Washington, D.C.: American Red Cross, 1929.
Anderson, Miriam. "Midwest Floods of 1993: St. Charles Experience." *Forum for
   Applied Research and Public Policy* 11 (1996): 128–130.
Andrews, William D. "The Literature of the 1727 New England Earthquake." *Early
   American Literature* 7 (1973): 281–294.
Bagnall, Norma Hayes. *On Shaky Ground: The New Madrid Earthquakes of 1811–1812.*
   Columbia: Univ. of Missouri Press, 1996.
Barnes, Jay. *Florida's Hurricane History.* Chapel Hill: Univ. of North Carolina
   Press, 1998.
Barry, John M. *Rising Tide: The Great Mississippi Flood of 1927 and How It Changed
   America.* New York: Simon and Schuster, 1997.
Bates, Charles C., and John F. Fuller. *America's Weather Warriors, 1814–1985.* College
   Station: Texas A&M Univ. Press, 1986.
Beck, Ulrich. *Risk Society: Towards a New Modernity.* Translated by Mark Ritter.
   London: Sage Publications, 1992.
Best, Gary Dean. *FDR and the Bonus Marchers, 1933–1935.* Westport, Conn.: Praeger,
   1992.
Biel, Steven. *Down with the Old Canoe: A Cultural History of the Titanic Disaster.* New
   York: Norton, 1996.
Bieringer, Paul, and Peter S. Ray. "A Comparison of Tornado Warning Lead Times with
   and without NEXRAD Doppler Radar." *Weather and Forecasting* 11 (1996): 47–52.
Bingham, Millicent Todd. "Miami: A Study in Urban Geography." *Tequesta* 8 (1948):
   73–107.
Birkland, Thomas A. "Factors Inhibiting a National Hurricane Policy." *Coastal
   Management* 25 (1997): 387–403.

Bolt, Bruce A. *Earthquakes and Geological Discovery*. New York: Scientific American Library, 1993.

Branner, J.C. "Earthquakes and Structural Engineering." *Bulletin of the Seismological Society of America* 3 (1913): 1–5.

Bruce, James P. "Natural Disaster Reduction and Global Change." *Bulletin of the American Meteorological Society* 75 (1994): 1831–1835.

Bruun, Per, and John M. De Grove. *Bayfill and Bulkhead Line Problems—Engineering and Management Considerations*. Studies in Public Administration, no. 18. Gainesville: Public Administration Clearing Service, Univ. of Florida, 1959.

Burby, Raymond J. *Sharing Environmental Risks: How to Control Governments' Losses in Natural Disasters*. Boulder, Colo.: Westview Press, 1991.

Burnham, John C. "A Neglected Field: The History of Natural Disasters." *Perspectives* (American Historical Association newsletter), Apr. 1988, 22–24.

Carson, Ruby Leach. "Forty Years of Miami Beach." *Tequesta* 15 (1955): 3–27.

Carter, Luther J. *The Florida Experience: Land and Water Policy in a Growth State*. Baltimore: Johns Hopkins Univ. Press, 1974.

———. "Weather Modification: Colorado Heeds Voters in Valley Dispute." *Science*, June 29, 1973.

Changnon, Stanley A., ed. *The Great Flood of 1993: Causes, Impacts, and Responses*. Boulder, Colo.: Westview Press, 1996.

Changnon, Stanley A., Kenneth E. Kunkel, and Beth C. Reinke. "Impacts and Responses to the 1995 Heat Wave: A Call to Action." *Bulletin of the American Meteorological Society* 77 (1996): 1497–1506.

Clark, Charles Edwin. "Science, Reason, and an Angry God: The Literature of an Earthquake." *New England Quarterly* 38 (1965): 340–362.

Clark, E. Culpepper. *Francis Warrington Dawson and the Politics of Restoration: South Carolina, 1874–1889*. University, Ala.: Univ. of Alabama Press, 1980.

Clarke, Thurston. *California Fault: Searching for the Spirit of a State Along the San Andreas*. New York: Ballantine, 1996.

Comerio, Mary C. *Disaster Hits Home: New Policy for Urban Housing Recovery*. Berkeley: Univ. of California Press, 1998.

Courtenay, William Ashmead. *Charleston Cyclone and Earthquake Scrapbooks*. Charleston: South Carolina Historical Society, 1885–1887.

Creel, Margaret Washington. *"A Peculiar People": Slave Religion and Community-Culture Among the Gullahs*. New York: New York Univ. Press, 1988.

Dacy, Douglas C., and Howard Kunreuther. *The Economics of Natural Disasters: Implications for Federal Policy*. New York: Free Press, 1969.

Daniel, Pete. *Deep'n As It Come: The 1927 Mississippi River Flood*. New York: Oxford Univ. Press, 1977.

Davidson, Keay. *Twister: The Science of Tornadoes and the Making of an Adventure Movie*. New York: Pocket Books, 1996.

Davis, Mike. *Ecology of Fear: Los Angeles and the Imagination of Disaster*. New York: Metropolitan Books, 1998.

Davison, Charles. *The Founders of Seismology*. Cambridge: Cambridge Univ. Press, 1927.

de Wetering, Maxine Van. "Moralizing in Puritan Natural Science: Mysteriousness in Earthquake Sermons." *Journal of the History of Ideas* 43 (1982): 417–438.

Dean, Dennis R. "The San Francisco Earthquake of 1906." *Annals of Science* 50 (1993): 501–521.

Doehring, Fred, Iver W. Duedall, and John M. Williams. *Florida Hurricanes and Tropical Storms: 1871–1993: An Historical Survey.* Technical Paper no. 71. Gainesville: Florida Sea Grant College Program, 1994.

Donoghue, Edmund R., Michael A. Graham, Jeffrey M. Jentzen, Barry D. Lifschultz, James L. Luke, and Haresh G. Mirchandani. "Criteria for the Diagnosis of Heat-Related Deaths: National Association of Medical Examiners." *American Journal of Forensic Medicine and Pathology* 18 (1997): 11–14.

Douglas, Mary, and Aaron Wildavsky. *Risk and Culture: An Essay on the Selection of Technological and Environmental Dangers.* Berkeley: Univ. of California Press, 1982.

Doyle, Don H. *New Men, New Cities, New South: Atlanta, Nashville, Charleston, Mobile, 1860–1910.* Chapel Hill: Univ. of North Carolina Press, 1990.

Dynes, Russell R., and Daniel Yutzy. "The Religious Interpretation of Disaster." *Topic* 5 (1965): 34–48.

Eliade, Mircea. *The Myth of the Eternal Return or, Cosmos and History.* Translated by Willard R. Trask. Princeton, N.J.: Princeton Univ. Press, 1954.

Ellis, F.P. "Mortality from Heat Illness and Heat-Aggravated Illness in the United States." *Environmental Research* 5 (1972): 1–58.

Engle, H.M. "The Earthquake Resistance of Buildings from the Underwriters' Point of View." *Bulletin of the Seismological Society of America* 19 (1929): 86–95.

Erikson, Kai T. *Everything in Its Path: Destruction of Community in the Buffalo Creek Flood.* New York: Simon and Schuster, 1976.

Farhar, Barbara C., and Julia Mewes. *Social Acceptance of Weather Modification: The Emergent South Dakota Controversy.* Institute of Behavioral Science, Program on Technology, Environment, and Man, monograph no. 23. Boulder: Univ. of Colorado, 1976.

Finley, John P. *Tornadoes: What They Are and How to Observe Them; With Practical Suggestions for the Protection of Life and Property.* New York: Insurance Monitor, 1887.

Fleagle, Robert G., ed. *Weather Modification: Science and Public Policy.* Seattle: Univ. of Washington, 1969.

Fleming, James Rodger. *Meteorology in America, 1800–1870.* Baltimore: Johns Hopkins Univ. Press, 1990.

Foxcroft, Thomas. *The Voice of the Lord, From the Deep Places of the Earth.* Boston: S. Gerrish, 1727.

Frank, Deborah A., Nicole Roos, Alan Meyers, Maria Napoleone, Karen Peterson, Amanda Cather, and L. Adrienne Cupples. "Seasonal Variation in Weight-for-Age in a Pediatric Emergency Room." *Public Health Reports* 111 (1996): 366–371.

Fraser, Walter J., Jr. *Charleston! Charleston!: The History of the Southern City.* Columbia: Univ. of South Carolina Press, 1989.

Freeman, John Ripley. *Earthquake Damage and Earthquake Insurance.* New York: McGraw Hill, 1932.

French, Jean, Roy Ing, Stephen Von Allmen, and Richard Wood. "Mortality from Flash Floods: A Review of National Weather Service Reports, 1969–81." *Public Health Reports* 98 (1983): 584–588.

Frey, Louis, and J. Richard Knop. "The Imperative Need for Uniform Mobile Home Safety Standards." *Washington and Lee Law Review* 30 (1973): 459–485.

Fried, John J. *Life Along the San Andreas Fault.* New York: Saturday Review Press, 1973.

Galbraith, John Kenneth. *The Great Crash 1929.* 1955. Reprint. New York: Avon, 1979.

Galway, Joseph G. "Early Severe Thunderstorm Forecasting and Research by the United States Weather Bureau." *Weather and Forecasting* 7 (1992): 564–587.

———. "Ten Famous Tornado Outbreaks." *Weatherwise* 34 (1981): 100–109.

Gere, James M., and Haresh C. Shah. *Terra Non Firma: Understanding and Preparing for Earthquakes.* San Francisco: W.H. Freeman, 1984.

Gilbert, G.K. "Earthquake Forecasts." *Science,* Jan. 22, 1909.

Glass, Roger I., Robert B. Craven, Dennis J. Bregman, Barbara J. Stoll, Neil Horowitz, Peter Kerndt, and Joe Winkle. "Injuries from the Wichita Falls Tornado: Implications for Prevention." *Science,* Feb. 15, 1980.

Grazulis, Thomas P. *Significant Tornadoes, 1680–1991.* St. Johnsbury, Vt.: Environmental Films, 1993.

Griswold, Oliver. *The Florida Keys and the Coral Reef.* Miami: Graywood Press, 1965.

Hall, David D. *Worlds of Wonder, Days of Judgment: Popular Religious Belief in Early New England.* New York: Knopf, 1989.

Halper, Louise A. "A New View of Regulatory Takings?" *Environment,* Jan./Feb. 1994.

Hanna, Alfred Jackson, and Kathryn Abbey Hanna. *Lake Okeechobee: Wellspring of the Everglades.* Indianapolis: Bobbs-Merrill, 1948.

Hansen, Gladys, comp. *Who Perished: A List of Persons Who Died as a Result of the Great Earthquake and Fire in San Francisco on April 18, 1906.* San Francisco: San Francisco Archives, 1980.

Hansen, Gladys, and Emmet Condon. *Denial of Disaster: The Untold Story and Photographs of the San Francisco Earthquake and Fire of 1906.* San Francisco: Robert A. Cameron, 1989.

Heppenheimer, T.A. *The Coming Quake: Science and Trembling on the California Earthquake Frontier.* New York: Times Books, 1988.

Hewitt, Kenneth. *Regions of Risk: A Geographical Introduction to Disaster.* Essex, U.K.: Longman, 1997.

———, ed. *Interpretations of Calamity: From the Viewpoint of Human Ecology.* Boston: Allen and Unwin, 1983.

History of St. Charles, Montgomery, and Warren Counties, Missouri....1885. Reprint. St. Louis: National Historical Co., 1969.

Hoffman, Frederick L. *Earthquake Hazards and Insurance.* Chicago: Spectator Co., 1928.

Hofstadter, Richard. *The Paranoid Style in American Politics and Other Essays.* New York: Knopf, 1965.

Horkheimer, Max. *Eclipse of Reason.* New York: Oxford Univ. Press, 1947.

Hosler, C.L. "Overt Weather Modification." *Reviews of Geophysics and Space Physics* 12 (1974): 523–527.

Howard, R. A., J. E. Matheson, and D. W. North. "The Decision to Seed Hurricanes."
    Science, June 16, 1972.

Howell, Wallace E. "Cloud Seeding and the Law in the Blue Ridge Area." *Bulletin of
    the American Meteorological Society* 46 (1965): 328–332.

Iacopi, Robert L. *Earthquake Country*. 4th ed. Tucson, Ariz.: Fisher Books, 1996.

The Impact of Catastrophes on Property Insurance. New York: Insurance Services
    Office, 1994.

Issel, William, and Robert W. Cherny. *San Francisco, 1865–1932: Politics, Power, and
    Urban Development*. Berkeley: Univ. of California Press, 1986.

Jenkins, Wilbert L. *Seizing the New Day: African Americans in Post–Civil War
    Charleston*. Bloomington: Indiana Univ. Press, 1998.

Jennings, Charles W. "New Geologic Map of California: A Summation of 140 Years of
    Geologic Mapping." *California Geology* 31 (1978): 77–96.

Jones, E.L. *The European Miracle: Environments, Economics and Geopolitics in the
    History of Europe and Asia*. Cambridge: Cambridge Univ. Press, 1981.

Jones, T. Stephen, Arthur P. Liang, Edwin M. Kilbourne, Marie R. Griffin, Peter A.
    Patriarca, Steven G. Fite Wassilak, Robert J. Mullan, Robert F. Herrick, H. Denny
    Donnell Jr., Keewhan Choi, and Stephen B. Thacker. "Morbidity and Mortality
    Associated with the July 1980 Heat Wave in St. Louis and Kansas City, Mo." *Journal
    of the American Medical Association* 247 (1982): 3327–3331.

Joyner, Charles. *Down by the Riverside: A South Carolina Slave Community*. Urbana:
    Univ. of Illinois Press, 1984.

Kahn, Judd. *Imperial San Francisco: Politics and Planning in an American City,
    1897–1906*. Lincoln: Univ. of Nebraska Press, 1979.

Kasson, John F. *Amusing the Million: Coney Island at the Turn of the Century*. New York:
    Hill and Wang, 1978.

Kelley, Robin D.G. *Race Rebels: Culture, Politics, and the Black Working Class*. New
    York: Free Press, 1994.

Kerr, Richard A. "Cloud Seeding: One Success in 35 Years." *Science*, Aug. 6, 1982.

Kreps, G.A. "Sociological Inquiry and Disaster Research." *Annual Review of Sociology* 10
    (1984): 309–330.

Landsea, Christopher W., Neville Nicholls, William M. Gray, and Lixion A. Avila.
    "Downward Trends in the Frequency of Intense Atlantic Hurricanes During the Past
    Five Decades." *Geophysical Research Letters* 23 (1996): 1697–1700.

Lantis, David W., Rodney Steiner, and Arthur E. Karinen. *California: Land of Contrast*.
    Belmont, Calif.: Wadsworth, 1963.

Lawson, Andrew C. *The California Earthquake of April 18, 1906: Report of the State
    Earthquake Investigation Commission*. Washington, D.C.: Carnegie Institution,
    1908.

———. "Seismology in the United States." *Bulletin of the Seismological Society of
    America* 1 (1911): 1–4.

Lennon, Gered, William J. Neal, David M. Bush, Orrin H. Pilkey, Matthew Stutz, and
    Jane Bullock. *Living With the South Carolina Coast*. Durham, N.C.: Duke Univ.
    Press, 1996.

Lewis, Oscar. *George Davidson: A Pioneer West Coast Scientist.* Berkeley: Univ. of California Press, 1954.

Lucas, David. *Lucas vs. The Green Machine.* Alexander, N.C.: Alexander Books, 1995.

Ludlum, David M. *The American Weather Book.* Boston: Houghton Mifflin, 1982.

———. *Early American Hurricanes, 1492–1870.* Boston: American Meteorological Society, 1963.

———. *Early American Tornadoes, 1586–1870.* Boston: American Meteorological Society, 1970.

———. "The Great Tornado Outbreak on 19 February 1884." *Weatherwise* 28 (1975): 84–87.

Machta, L., and D.L. Harris, "Effects of Atomic Explosions on Weather." *Science,* Jan. 21, 1955.

Maloney, Frank E., Sheldon J. Plager, and Fletcher N. Baldwin Jr. *Water Law and Administration: The Florida Experience.* Gainesville: Univ. of Florida Press, 1968.

Manwaring, C.T. "Report of Committee on Building For Safety Against Earthquakes: Preliminary Report on Guarding Against Panic." *Bulletin of the Seismological Society of America* 15 (1925): 213–221.

May, Elaine Tyler. *Homeward Bound: American Families in the Cold War Era.* New York: Basic Books, 1988.

May, Peter J. *Recovering from Catastrophes: Federal Disaster Relief Policy and Politics.* Westport, Conn.: Greenwood, 1985.

McCullough, David. *The Johnstown Flood: The Incredible Story Behind One of the Most Devastating "Natural" Disasters America Has Ever Known.* New York: Simon and Schuster, 1968.

McDonald, James R., and John F. Mehnert. "Review of Standard Practice for Wind-Resistant Manufactured Housing." *Journal of Aerospace Engineering* 2 (1989): 88–96.

McPhee, John. *Assembling California.* New York: Farrar, Straus and Giroux, 1993.

Meltsner, Arnold J. "The Communication of Scientific Information to the Wider Public: The Case of Seismology in California." *Minerva* 17 (1979): 331–354.

Meyer, William B. "Urban Heat Island and Urban Health: Early American Perspectives." *Professional Geographer* 43 (1991): 38–48.

Monmonier, Mark. *Cartographies of Danger: Mapping Hazards in America.* Chicago: Univ. of Chicago Press, 1997.

Morris, Edward A. "The Law and Weather Modification." *Bulletin of the American Meteorological Society* 46 (1965): 618–622.

Munroe, Ralph Middleton, and Vincent Gilpin. *The Commodore's Story.* N.p.: Ives Washburn, 1930.

National Research Council. *A Safer Future: Reducing the Impacts of Natural Disasters.* Washington, D.C.: National Academy Press, 1991.

———. *Severe Storms: Prediction, Detection, and Warning.* Washington, D.C.: National Academy of Sciences, 1977.

———. *Technological and Scientific Opportunities for Improved Weather and Hydrological Services in the Coming Decade.* Washington, D.C.: National Academy of Sciences, 1980.

―――. *Toward a New National Weather Service—Assessment of NEXRAD Coverage and Associated Weather Services*. Washington, D.C.: National Academy Press, 1995.

―――. *Weather and Climate Modification: Problems and Progress*. Washington, D.C.: National Academy of Sciences, 1973.

O'Connor, Charles J., Francis M. McLean, Helen Swett Artieda, James Marvin Motley, Jessica Peixotto, and Mary Roberts Coolidge, comps. *San Francisco Relief Survey: The Organization and Methods of Relief Used After the Earthquake and Fire of April 18, 1906*. New York: Survey Associates, 1913.

Obasi, G.O.P. "WMO's Role in the International Decade for Natural Disaster Reduction." *Bulletin of the American Meteorological Society* 75 (1994): 1655–1661.

Officer, Charles, and Jake Page. *Tales of the Earth: Paroxysms and Perturbations of the Blue Planet*. New York: Oxford Univ. Press, 1993.

Official Building Laws, City and County of San Francisco. San Francisco: Daily Pacific Builder, 1921.

Oliver-Smith, Anthony. "Disaster Context and Causation: An Overview of Changing Perspectives in Disaster Research." In *Natural Disasters and Cultural Responses*, edited by Anthony Oliver-Smith. Publication 36 of *Studies in Third World Societies* (1986): 1–34.

O'Malley, Michael. *Keeping Watch: A History of American Time*. New York: Viking, 1990.

Ortner, Sherry B. *Making Gender: The Politics and Erotics of Culture*. Boston: Beacon, 1996.

Palm, Risa, and Michael E. Hodgson. *After a California Earthquake: Attitude and Behavior Change*. Chicago: Univ. of Chicago Press, 1992.

Peacock, Walter Gillis, Betty Hearn Morrow, and Hugh Gladwin. *Hurricane Andrew: Ethnicity, Gender, and the Sociology of Disasters*. London: Routledge, 1997.

Penick, James Lal, Jr., *The New Madrid Earthquakes*. Rev. ed. Columbia: Univ. of Missouri Press, 1981.

Pielke, Roger A., Jr. *Hurricane Andrew in South Florida: Mesoscale Weather and Societal Responses*. Boulder, Colo.: National Center for Atmospheric Research, 1995.

―――. "Reframing the U.S. Hurricane Problem." *Society and Natural Resources* 10 (1997): 485–499.

Pilat, Oliver, and Jo Ranson. *Sodom by the Sea: An Affectionate History of Coney Island*. Garden City, N.Y.: Doubleday, 1941.

Pilkey, Orrin H., and Katharine L. Dixon. *The Corps and the Shore*. Washington, D.C.: Island Press, 1996.

Pilkey, Orrin H. Jr., Dinesh C. Sharma, Harold R. Wanless, Larry J. Doyle, Orrin H. Pilkey Sr., William J. Neal, and Barbara L. Gruver. *Living with the East Florida Shore*. Durham, N.C.: Duke Univ. Press, 1984.

Platt, Rutherford. "Hazard Mitigation: Cornerstone or Grains of Sand?" *Natural Hazards Observer* 21 (1996): 10–11.

Platt, Rutherford H. *Disasters and Democracy: The Politics of Extreme Natural Events*. Washington, D.C.: Island Press, 1999.

―――. "Life After Lucas: The Supreme Court and the Downtrodden Coastal Developer." *Natural Hazards Observer* 17 (1992): 8–9.

Pound, Arthur. "Conquering the Earthquake Crisis." *Independent,* July 25, 1925.

Powell, J.W. "Our Recent Floods." *North American Review* 155 (1892): 149–159.

Powers, Bernard E., Jr. *Black Charlestonians: A Social History, 1822–1885.* Fayetteville: Univ. of Arkansas Press, 1994.

Prescott, William H. "Circumstances Surrounding the Preparation and Suppression of a Report on the 1868 California Earthquake." *Bulletin of the Seismological Society of America* 72 (1982): 2389–2393.

Prince, Samuel Henry. *Catastrophe and Social Change.* New York: Columbia Univ. Press, 1920.

Quarantelli, E.L., and Russell R. Dynes. "Response to Social Crisis and Disaster." *Annual Review of Sociology* 3 (1977): 23–49.

Rahn, Perry H. "Lessons Learned from the June 9, 1972, Flood in Rapid City, South Dakota." *Bulletin of the Association of Engineering Geologists* 12 (1975): 83–97.

Rangno, Arthur L., and Peter V. Hobbs. "A New Look at the Israel Cloud Seeding Experiments." *Journal of Applied Meteorology* 34 (1995): 1169–1193.

Redford, Polly. *Billion-Dollar Sandbar: A Biography of Miami Beach.* New York: Dutton, 1970.

Reese, Joe Hugh. *Florida's Great Hurricane.* Miami: Lysle E. Fesler, 1926.

Richard, Melvin J. "Tidelands and Riparian Rights in Florida." *Miami Law Quarterly* 3 (1949): 339–364.

Ross, Andrew. *Strange Weather: Culture, Science, and Technology in the Age of Limits.* London: Verso, 1991.

"The Row over Florida Relief." *Literary Digest.* Oct. 16, 1926.

Sargent, Frederick. "A Dangerous Game: Taming the Weather." *Bulletin of the American Meteorological Society* 48 (1967): 452–458.

Schaefer, Vincent J. "A Call for Action." *Journal of Weather Modification* 2 (1970): 1–13.

Scheer, Robert. *With Enough Shovels: Reagan, Bush, and Nuclear War.* New York: Random House, 1982.

Segall, Paul. "New Insights into Old Earthquakes." *Nature,* July 10, 1997.

Semenza, Jan C., Carol H. Rubin, Kenneth H. Falter, Joel D. Selanikio, W. Dana Flanders, Holly L. Howe, and John L. Wilhelm. "Heat-Related Deaths During the July 1995 Heat Wave in Chicago." *New England Journal of Medicine* 335 (1996): 84–90.

Sewell, W.R. Derrick, ed. *Human Dimensions of Weather Modification.* Chicago: Univ. of Chicago Press, 1966.

Shumate, Albert. *The California of George Gordon and the 1849 Sea Voyages of His California Association.* Glendale, Calif.: Arthur H. Clark, 1976.

Sirkin, Alan. "Engineering Overview of Hurricane Andrew in South Florida." *Journal of Urban Planning and Development* 121 (1995): 1–10.

Smiley, Nixon. *Knights of the Fourth Estate: The Story of the* Miami Herald. Miami: E.A. Seeman, 1974.

Sofen, Edward. *The Miami Metropolitan Experiment.* Bloomington: Indiana Univ. Press, 1963.

Southern Pacific Company. *San Francisco Imperishable.* San Francisco: Southern Pacific Company, 1906.

Sparks, Peter R. "Development of the South Carolina Coast 1959–1989: Prelude to a Disaster." In *Hurricane Hugo One Year Later*, edited by Benjamin L. Sill and Peter R. Sparks. New York: American Society of Civil Engineers, 1991.

———. "The Facts About Hurricane Hugo—What It Was, What It Wasn't and Why It Caused So Much Damage." In *Hurricane Hugo One Year Later*, edited by Benjamin L. Sill and Peter R. Sparks. New York: American Society of Civil Engineers, 1991.

Spence, Clark C. *The Rainmakers: American 'Pluviculture' to World War II*. Lincoln: Univ. of Nebraska Press, 1980.

St.-Amand, Pierre, Ray J. Davis, and Robert D. Elliott. "Report on Rapid City Flood of June 9, 1973." *Journal of Weather Modification* 5 (1973): 318–346.

Steinberg, Theodore. *Slide Mountain, Or the Folly of Owning Nature*. Berkeley: Univ. of California Press, 1995.

Steinbrugge, Karl V. *Earthquake Hazard in the San Francisco Bay Area: A Continuing Problem in Public Policy*. Berkeley: Institute of Governmental Studies, Univ. of California, 1968.

———. *Earthquakes, Volcanoes, and Tsunamis: An Anatomy of Hazards*. New York: Skandia America Group, 1982.

Stetson, John. "Message from the President." *Florida Architect*, Feb. 1960.

Stevens, Walter B. *Centennial History of Missouri: One Hundred Years in the Union 1820–1921*. St. Louis: S.J. Clark Publishing Co., 1921.

Stewart, George R. *Storm*. New York: Random House, 1941.

Stockbridge, Frank Parker, and John Holliday Perry. *Florida in the Making*. Jacksonville, Fla.: de Bower Publishing Co., 1926.

Stockton, Robert P. *The Great Shock: The Effects of the 1886 Earthquake on the Built Environment of Charleston, South Carolina*. Easley, S.C.: Southern Historical Press, 1986.

Strickland, John Scott. "Traditional Culture and Moral Economy: Social and Economic Change in the South Carolina Low Country, 1865–1910." In *The Countryside in the Age of Capitalist Transformation: Essays in the Social History of Rural America*, edited by Steven Hahn and Jonathan Prude. Chapel Hill: Univ. of North Carolina Press, 1985.

Tebeau, Charlton W. *A History of Florida*. Coral Gables: Univ. of Miami Press, 1971.

Thomas, William A., ed. *Legal and Scientific Uncertainties of Weather Modification*. Durham, N.C.: Duke Univ. Press, 1977.

Tibbetts, John. "Everybody's Taking the Fifth." *Planning*, Jan. 1995.

Weather Modification Association. *Weather Modification: Some Facts About Seeding Clouds*. Fresno, Calif.: Weather Modification Association, 1977.

White, Gilbert F. "A Perspective on Reducing Losses from Natural Hazards." *Bulletin of the American Meteorological Society* 75 (1994): 1237–1240.

Whitman, Steven, Glenn Good, Edmund R. Donoghue, Nanette Benbow, Wenyuan Shou, and Shanxuan Mou. "Mortality in Chicago Attributed to the July 1995 Heat Wave." *American Journal of Public Health* 87 (1997): 1515–1518.

Whitnah, Donald R. *A History of the United States Weather Bureau*. Urbana: Univ. of Illinois Press, 1961.

Whitney, J.D. "Earthquakes." *North American Review* 108 (1869): 578–610.

Will, Lawrence E. *Okeechobee Hurricane and the Hoover Dike.* St. Petersburg, Fla.: Great Outdoors Publishing, 1961.

Willis, Bailey. "Earthquake Risk in California." *Bulletin of the Seismological Society of America* 14 (1924): 9–15.

Wollenberg, Charles. "Life on the Seismic Frontier: The Great San Francisco Earthquake." *California History* 71 (Winter 1992/1993): 494–509.

Worster, Donald. *Dust Bowl: The Southern Plains in the 1930s.* New York: Oxford Univ. Press, 1979.

———. *Rivers of Empire: Water, Aridity, and the Growth of the American West.* New York: Pantheon, 1985.

Zelizer, Viviana A. *The Social Meaning of Money.* New York: Basic Books, 1994.

Zoback, Mark D., and Mary Lou Zoback. "*In Situ* Stress, Crustal Strain, and Seismic Hazard Assessment in Eastern North America." In *Earthquake Hazards and the Design of Constructed Facilities in the Eastern United States,* edited by Klaus H. Jacob and Carl J. Turkstra. New York: New York Academy of Sciences, 1989.

GOVERNMENT PUBLICATIONS

Coffman, Jerry L., Carl A. von Hake, and Carl W. Stover, eds. *Earthquake History of the United States.* Boulder, Colo.: Department of Commerce, NOAA, and Department of Interior, U.S. Geological Survey, 1982.

Dobney, Fredrick J. *River Engineers on the Middle Mississippi: A History of the St. Louis District, U.S. Army Corps of Engineers.* Washington, D.C.: GPO, 1978.

Earthquake Hazards, Risk, and Mitigation in South Carolina and the Southeastern United States. Charleston: South Carolina Seismic Safety Consortium, 1986.

Executive Office of the President, *Sharing the Challenge: Floodplain Management into the 21st Century.* Report of the Interagency Floodplain Management Review Committee. Washington, D.C., 1994.

Floodplain Management in the United States: An Assessment Report. 2 vols. Prepared for the Federal Interagency Floodplain Management Task Force by L.R. Johnston Associates. Washington, D.C.: Federal Emergency Management Agency, 1992.

Florida Hurricane Report Concerning Hurricane "Donna." N.p., 1961.

Fuller, Myron L. *The New Madrid Earthquake.* U.S. Geological Survey Bulletin 494. Washington, D.C., 1912.

Greene, Majorie R., and Paula L. Gori. "Earthquake Hazards Information Dissemination: A Study of Charleston, South Carolina." Open File Report 82-233. Reston, Va.: U.S. Geological Survey, 1982.

Haeussner, Theodore E. "Tides and Flooding Incident to Winds of Hurricane Force." In Florida Hurricane Damage Study Committee, *Florida Hurricane Survey Report 1965.* N.p., n.d.

Hebert, Paul J., Jerry D. Jarrell, and Max Mayfield. *The Deadliest, Costliest, and Most Intense United States Hurricanes of This Century (And Other Frequently Requested*

Hurricane Facts). NOAA Technical Memorandum NWS TPC-1. Washington, D.C., 1996.

Message of Governor George C. Pardee to the Extra Session of the Legislature of California. June 2, 1906. Sacramento: State Printing Office, 1906.

Missouri State Emergency Management Agency. "Out of Harm's Way: The Missouri Buyout Program." N.p., n.d.

National Advisory Committee on Oceans and Atmosphere. *The Future of the Nation's Weather Services.* Special Report to the President and the Congress. Washington, D.C., 1982.

———. *The Nation's River and Flood Forecasting and Warning Service.* Special Report to the President and the Congress. Washington, D.C., 1983.

National Commission on Manufactured Housing. *Final Report.* Washington, D.C.: GPO, 1994.

Nau, J.M., and A.K. Gupta. "The Earthquake Threat and Its Mitigation in the Southeastern United States." In *A Workshop on "The 1886 Charleston, South Carolina, Earthquake and Its Implications for Today,"* edited and compiled by Walter W. Hays, Paula L. Gori, and Carla Kitzmiller. Reston, Va.: U.S. Geological Survey, 1983.

Peters, Kenneth E., and Robert B. Herrmann, eds. *First-Hand Observations of the Charleston Earthquake of August 31, 1886, and Other Earthquake Materials: Reports of W.J. McGee, Earle Sloan, Gabriel E. Manigault, Simon Newcomb, and Others.* Columbia: South Carolina Geological Survey, 1986.

U.S. Army Corps of Engineers. *The Great Flood of 1993: Post-Flood Report.* Main Report. Chicago: Corps North Central Division, 1994.

U.S. Congress. House. Committee on Appropriations. *Department of the Interior and Related Agencies Appropriations for 1998.* 105th Cong., 1st sess., 1997, pt. 8.

U.S. Congress. House. Committee on Banking, Finance, and Urban Affairs. *Housing Needs in Earthquake Disaster Areas.* 101st Cong., 2d sess., 1990.

———. *Manufactured Housing.* 102d Cong., 2d sess., 1992.

U.S. Congress. House. Committee on Government Operations. *Managing the Federal Government: A Decade of Decline.* 102d Cong., 2d sess., 1993.

U.S. Congress. House. Committee on Interstate and Foreign Commerce. *National Weather Warning System.* 94th Cong., 2d sess., 1976.

———. *Weather Modification and Early-Warning Systems.* 93d Cong., 2d sess., 1974.

U.S. Congress. House. Committee on Public Works. *Disaster Relief Act of 1965.* 89th Cong., 2d sess., 1966.

———. *South Dakota Flood Disaster.* 92d Cong., 2d sess., 1972.

U.S. Congress. House. Committee on Public Works and Transportation. *Reauthorization of the Federal Disaster Relief Program.* 100th Cong., 1st sess., 1987.

U.S. Congress. House. Committee on Science, Space, and Technology. *Affordable Housing and Construction R&D.* 103d Cong., 1st sess., 1993.

———. *NEXRAD, Tornado Warnings, and National Weather Service Modernization.* 103d Cong., 2d sess., 1994.

———. *Tornado Forecasting and Severe Storm Warning.* 100th Cong., 1st sess., 1989.

U.S. Congress. House. Committee on Science and Technology. *Earthquakes in the Eastern United States.* 98th Cong., 2d sess., 1984.

———. *Field Briefing on Tornado Prediction and Preparedness in the Carolinas During the Severe Weather of March 28, 1984.* 98th Cong., 2d sess., 1984.

———. *H.R. 13715—National Weather Service Act of 1978.* 95th Cong., 2d sess., 1978.

U.S. Congress. House. Committee on Small Business. *Federal Natural Disaster Assistance Programs.* 95th Cong., 1st sess., 1977.

U.S. Congress. House. Committee on World War Veterans' Legislation. *Florida Hurricane Disaster.* 74th Cong., 2d sess., 1935.

U.S. Congress. House. *Report on Mississippi River Urban Areas From Hampton, Illinois, to Mile 300.* 87th Cong., 2d sess., 1962, H. Doc. 564.

U.S. Congress. Senate. Committee on Appropriations. *Departments of State, Justice, and Commerce, the Judiciary, and Related Agencies Appropriations for Fiscal Year 1975.* 93d Cong., 2d sess., 1974.

———. *Departments of State, Justice, and Commerce, the Judiciary, and Related Agencies Appropriations for Fiscal Year 1978.* 95th Cong., 1st sess., 1977.

U.S. Congress. Senate. Committee on Banking, Housing, and Urban Affairs. *1973 Housing and Development Legislation.* 93d Cong., 1st sess., 1973.

———. *SBA Administration of the Disaster Relief Program.* 92d Cong., 2d sess., 1972.

U.S. Congress. Senate. Committee on Commerce, Science, and Transportation. *Weather Modification: Programs, Problems, Policy, and Potential.* 95th Cong., 2d sess., 1978.

U.S. Congress. Senate. Committee on Environment and Public Works. *Disaster Relief Act Amendment of 1980.* 96th Cong., 2d sess., 1980, S. Rept. no. 96-891.

———. *Disaster Relief Act Amendments of 1983.* 98th Cong., 1st sess., 1984.

———. *Disaster Relief Act Amendments of 1988.* 100th Cong., 2d sess., 1988.

U.S. Congress. Senate. Committee on Foreign Relations. *Weather Modification.* 93d Cong., 2d sess., 1974.

U.S. Congress. Senate. Committee on Governmental Affairs. *Nomination of Louis O. Giuffrida.* 97th Cong., 1st sess., 1981.

U.S. Congress. Senate. Committee on Public Works. *Federal Response to Hurricane Camille.* 91st Cong., 2d sess., 1970, pts. 1–5.

———. *To Investigate the Adequacy and Effectiveness of Federal Disaster Relief Legislation.* 93d Cong., 1st sess., 1973, pt. 3.

U.S. Congress. Senate. *Federal Disaster Assistance: Report of the Senate Task Force on Funding Disaster Relief.* 104th Cong., 1st sess., 1995, S. Doc. 104-4.

U.S. Department of Commerce. NOAA. *Big Thompson Canyon Flash Flood of July 31–August 1, 1976.* Natural Disaster Survey Report no. 76-1. Rockville, Md., 1976.

———. *Black Hills Flood of June 9, 1972.* Natural Disaster Survey Report no. 72-1. Rockville, Md., 1972.

———. *Johnstown, Pennsylvania, Flash Flood of July 19–20, 1977.* Natural Disaster Survey Report no. 77-1. Rockville, Md., 1977.

———. *July 1995 Heat Wave.* Natural Disaster Survey Report, no number. Silver Spring, Md., 1995.

————. *Ohio River Valley Flood of March 1997.* Office of Hydrology Service
  Assessment. Silver Spring, Md., 1998.
————. *The Widespread Tornado Outbreak of April 3–4, 1974.* Natural Disaster Survey
  Report no. 74-1. Rockville, Md., 1974.
U.S. Department of Defense and Department of Commerce, NOAA. *Project
  Stormfury: Annual Report 1972.* Miami, Fla., 1973.
U.S. General Accounting Office. *Federal Disaster Assistance: What Should the Policy Be?*
  PAD-80-39. Washington, D.C., 1980.
————. *Weather Forecasting: Cost Growth and Delays in Billion-Dollar Weather Service
  Modernization.* GAO/IMTEC-92-12FS. Washington, D.C., 1991.
U.S. Office of Emergency Preparedness. *Disaster Preparedness.* Report to Congress.
  Washington, D.C., 1972.
Wahl, Kenneth L., Wilbert O. Thomas Jr., and Robert M. Hirsch. "The Stream-
  Gaging Program of the U.S. Geological Survey." U.S. Geological Survey Circular
  no. 1123. Reston, Va., 1995.
Wallace, Robert E., ed., *San Andreas Fault System, California.* U.S. Geological Survey
  Professional Paper 1515. Washington, D.C.: GPO, 1990.
Wentworth, Carl M. "The Changing Tectonic Basis for Regulatory Treatment of the
  1886 Charleston, South Carolina, Earthquake in the Design of Power Reactors." In
  *A Workshop on "The 1886 Charleston, South Carolina, Earthquake and Its Implications
  for Today,"* edited and compiled by Walter W. Hays, Paula L. Gori, and Carla
  Kitzmiller. Reston, Va.: U.S. Geological Survey, 1983.

NEWSPAPERS

Atlanta Journal and Constitution, 1990.
Baltimore Sun, 1886.
Belle Glade News (Canal Point, Fla.), 1928.
Cincinnati Enquirer, 1997.
Clearwater (Fla.) Sun and Herald, 1928.
Daily Picayune (New Orleans), 1893, 1896.
Denver Post, 1968, 1972, 1973.
Durham (N.C.) Morning Herald, 1989.
Everglades News (Canal Point, Fla.), 1926, 1928.
Fresno (Calif.) Bee, 1994.
Fulton County (Pa.) News, 1964.
Fulton Democrat (Fulton County, Pa.), 1964, 1965.
Gettysburg (Pa.) Times, 1977.
Hannibal (Mo.) Courier-Post, 1993.
Homestead (Fla.) Enterprise, 1926.
Huron (S.Dak.) Plainsman, 1993.
Ledger-Enquirer (Columbus, Ga.), 1995, 1996, 1997.
Los Angeles Times, 1972.
Louisville Courier-Journal, 1997, 1998.

Mercersburg (Pa.) Journal, 1962, 1965.

Miami Daily News, 1926, 1928, 1935.

Miami Daily Tribune, 1935.

Miami Herald, 1926, 1928, 1935.

New York Evening Post, 1893, 1926.

New York Herald, 1886.

New York Times, 1886, 1906, 1926, 1928, 1936, 1954, 1970, 1978, 1992–1999.

New York Tribune, 1886, 1896, 1902.

News and Courier (Charleston, S.C.), 1885, 1886.

Orlando (Fla.) Sentinel, 1994.

Philadelphia Inquirer, 1978.

Public Opinion, (Chambersburg, Pa.), 1976.

Rapid City (S.Dak.) Journal, 1972, 1982.

San Francisco Bulletin, 1906.

San Francisco Call, 1906.

San Francisco Chronicle, 1868, 1906, 1965, 1995.

San Francisco Examiner, 1906.

St. Charles (Mo.) Banner-News, 1971, 1973, 1975, 1977, 1978.

St. Charles (Mo.) Journal, 1958, 1971, 1978, 1987.

St. Charles (Mo.) Post, 1982, 1996.

St. Louis Globe-Democrat, 1971, 1973.

St. Louis Post-Dispatch, 1896, 1903, 1983–1986, 1988, 1990–1994, 1996.

St. Louis Republic, 1896.

St. Petersburg (Fla.) Times, 1992, 1994.

State (Columbia, S.C.), 1988.

Tampa (Fla.) Tribune, 1926, 1996.

Virginian-Pilot and the Ledger-Star (Norfolk, Va.), 1989.

Wall Street Journal, 1926, 1928.

Washington Post, 1926.

INTERVIEWS CONDUCTED BY THE AUTHOR

Anderson, Miriam G. Telephone interview, Dec. 12, 1996.

Benton, George. Telephone interview, Dec. 17, 1997.

Ehlmann, Steve E. Telephone interview, Dec. 12, 1996.

Hammond, David. Telephone interview, Nov. 25, 1996.

Harris-Wheeler, Shelia. Telephone interview, Dec. 4, 1996.

Hirn, Richard. Cleveland, Apr. 11, 1997.

Kalkstein, Laurence S. Telephone interview, Mar. 26, 1999.

Katt, Buck. Telephone interview, Dec. 5, 1996.

Lauer, Steven G. Telephone interview, Nov. 22, 1996.

Mahan, Miriam. Telephone interview, Dec. 4, 1996.

May, Darren L. Telephone interview, Dec. 17, 1996.

McCabe, Ron. Telephone interview, Nov. 27, 1996.

Reidt, Jody. Telephone interview, Nov. 22, 1996.

Szilasi, Tom. Telephone interview, Dec. 4, 1996.

White, Herb. Telephone interview, May 22, 1997.

Wilhelm, John. Telephone interview, Mar. 30, 1999.

UNPUBLISHED DISSERTATIONS

Geschwind, Carl-Henry. "Earthquakes and Their Interpretation: The Campaign for Seismic Safety in California, 1906–1933." Ph.D. diss., Johns Hopkins Univ., 1996.

Hoff, Judy Lee. "A Water Balance Evaluation of the Effects of Climate Variability and Human Modification on the Flow Regime of the Mississippi." Ph.D. diss., Louisiana State Univ., 1994.

Linehan, Urban J. "Tornado Deaths in the United States." Ph.D. diss., Clark Univ., 1955.

Sessa, Frank Bowman. "Real Estate Expansion and Boom in Miami and Its Environs During the 1920s." Ph.D. diss., Univ. of Pittsburgh, 1950.

Shute, Michael Nathaniel. "Earthquakes and Early American Imagination: 'Decline and Renewal in Eighteenth-Century Puritan Culture.'" Ph.D. diss., Univ. of California, Berkeley, 1977.

INDEX

....